METAL CONTAMINATION OF FOOD

Second Edition

METAL CONTAMINATION OF FOOD

Second Edition

CONOR REILLY

B.Sc., B.Phil., H.dip.Ed., Ph.D., FAIFST

Professor and Head of School,
School of Public Health,
Queensland University of Technology, Brisbane, Australia

ELSEVIER APPLIED SCIENCE
LONDON and NEW YORK

ELSEVIER SCIENCE PUBLISHERS LTD
Crown House, Linton Road, Barking, Essex IG11 8JU, England

Sole Distributor in the USA and Canada
ELSEVIER SCIENCE PUBLISHING CO., INC.
655 Avenue of the Americas, New York, NY 10010, USA

First edition 1980
Second edition 1991

WITH 11 TABLES

© 1991 ELSEVIER SCIENCE PUBLISHERS LTD

British Library Cataloguing in Publication Data

Reilly, Conor
    Metal contamination of food—2nd ed.
    1. Food. Contamination by metal compounds
    I. Title
    363.192

    ISBN 1-85166-540-4

Library of Congress Cataloging-in-Publication Data

Reilly, Conor.
    Metal contamination of food/Conor Reilly—2nd ed.
        p.   cm.
    Includes bibliographical references and index.
    ISBN 1-85166-540-4
    1. Metals—Analysis.   2. Food—Analysis.   3. Food contamination.
    I. Title.
    TX571.M48R45   1991
    363.19'2—dc20
                                                              90-42142
                                                                  CIP

Printed in Great Britain at the University Press, Cambridge

PREFACE TO THE SECOND EDITION

It is a brave publisher who will launch a new technical journal or a new book on to what must appear to be, to all but the most optimistic, an already saturated market. Not only do scientists groan at the continuous flood of publications with which they feel obliged to keep abreast, but they must pay attention to shrinking budgets which demand that a new periodical for the library be paid for by the cancellation of one already on the shelves. Prasad, writing in the editorial of the first issue of a new trace element journal[1], sought to justify the venture on the grounds that there was a need for a single publication that would bring together data currently scattered in many journals, making it difficult for researchers to keep trace of significant advances which were taking place so rapidly. He expressed the hope that his readers would benefit greatly from the journal's comprehensive overview of the many facets of trace element research.

I used a somewhat similar justification in the Preface of the first edition of this book. My aim, I wrote, was to present a brief and useful summary of available information gathered from many sources and thus to make available in one place what otherwise might require time-consuming literature searches.

Now, 10 years later, I make the same claim, that this is a useful, compact source of information that otherwise would not be readily available to the hard-pressed scientist and to others who do not have easy access to technical databases. The need for a new edition is reinforced by the fact that, during that decade, there have been several significant developments which have made the first edition no longer as useful as it was meant to be to its users.

The continuing flow of new data on metals in food and on their technical and health implications has made some of the earlier information obsolescent. Of major importance have been a variety of improvements in analytical equipment and techniques, notably the introduction of inductively coupled plasma mass spectrometry (ICP-MS) as well as of Zeeman mode atomic absorption spectrophotometry and the availability of second-generation certified reference materials (CRMs). Perhaps no less significant than the technical developments has been a wider acceptance of the need for greater care and the practise of immaculate laboratory hygiene when dealing with trace levels of metals in food.

Another development of significance has already been referred to, namely the appearance of a number of new periodicals dealing with aspects of metals in foods and their relation to health. Several of these, such as *Food Additives and Contaminants* (from 1984), *Journal of Micronutrient Analysis* (from 1985), *Journal of Trace Elements and Electrolytes in Health and Disease* (from 1987) and *Journal of Trace Elements in Experimental Medicine* (from 1988), have been of considerable use in the preparation of this volume.

Reviews of the first edition showed a surprisingly wide spread of interest in the subject of metals in food among scientists in different fields. They included physicians, environmental health officers, food manufacturers, technologists and public analysts. The book was recommended by one reviewer to food scientists, quality control professionals and regulatory officials and, as a library reference in schools of public health and environmental science. Another reviewer noted that it would be of value also to the interested lay reader.

It was pleasing for the author to find that the book was welcomed in many countries, because of the international nature of its subject and the worldwide problems that metal contamination can cause. A New Zealand reviewer noted that the material was presented in a way to which his fellow countrymen could easily relate. In Scotland it was recommended for adoption as a textbook for students preparing for the Mastership in Chemical Analysis. A Russian translation, published in 1985, testified to its perceived usefulness in the USSR.

It is hoped that the new edition will also appeal to a wide, international audience. To some extent the style used and selection of material for inclusion in the text have been determined by an expectation of a wide readership spectrum. Apologies are made to specialist readers for the over-simplifications and the generalisations which resulted from this approach. It is to be hoped that the disadvantages will be outweighed by the advantage of presenting a thorough overview of the field in a concise, readable and not too technical manner for less specialised readers from many areas of interest and from differing backgrounds.

My gratitude is due to many people for help in writing this book. My wife, Ann, my secretaries, the ever helpful information librarians at QUT, as well as my research associates and graduate students who shared in my enthusiasm for investigating trace metals in the diet, are all thanked for their contributions.

CONOR REILLY
Brisbane, Queensland, 23 March 1990

REFERENCES

1. Prasad, A. S. (1988). Editorial: Why a new trace element journal? *J. Trace Elem. Exp. Med.* 1, 1–2.

PREFACE TO THE FIRST EDITION

Some eighty of the hundred-odd elements of the Periodic Table are metals. Several of the metals are known to be essential for human life; others are toxic, even in small amounts. These metals occur in all foodstuffs, in greater or lesser amounts depending on various circumstances. The presence of metals in food can have both good and bad consequences. It can be of interest to food processors, nutritionists, toxicologists and a wide variety of other scientists. This book is concerned entirely with the presence of these metals in food. It concentrates on metals which are generally considered undesirable, at least in more than trace quantities, in food. In a general way it follows the pathways by which those metals get into food, and examines the significance—from the point of view of the manufacturer as well as of the scientist—of that contamination. It considers how man has sought to protect himself from the undesirable effects of metal contamination of food by passing laws and regulations. The international consequences of such laws are evidenced by the efforts of the United Nations and other organisations to develop a uniform and harmonised universal code of regulations concerning food.

In the second part of the study attention is given to individual metals which occur as food contaminants. A great deal of investigation has been carried out on many of these metals from biochemical, medical and other points of view. The findings of these studies have been published in many different journals and some are still only available in restricted form. An attempt has been made in this book to present a brief and useful summary of available information gathered from many sources. The aim has been to make available in one place what otherwise might require time-consuming literature searches.

Metals have served man well since he started on the long road of technological progress. We chart human progress by the metals we have used: thus we refer to the Bronze and Iron Ages. Only relatively recently have we begun to appreciate the significance of metals in a facet of human life besides that of technology. We now know a great deal about the part played by metals in the structure and function of the human body. We classify some metals among the essential nutrients. We know that others are destructive of human life and development. But we are also aware that certain metals can have both good and bad effects on the human organism, depending on the amount of metal present and other factors. Apart from the biological and medical scientist, the food chemist and technologist also have a major interest in the metals. They know the importance of metals in food not only with regard to health and toxicity, but also from the point of view of food quality, processing and commercial characteristics.

The general public is more aware today than ever before of the part metals in food play in people's lives. They are exposed to advertising extolling the nutrient value of minerals in food and in health supplements of many kinds. They are alert to the dangers of heavy metal contamination of what they eat and look to legislation to protect themselves against an excess of undesirable metals in their diet.

It is with these diverse groups of people in mind that this book was written. They all have a commendable interest in the topic, but information has not always been readily available to meet their needs. True, the specialist will frequently want to delve deeply into the question and seek out original reports, but often the intelligent layman, as well as the technician and specialist, will want a comprehensive, general treatment of the subject such as is given here.

Metal contamination of food is by no means confined to any one nation. Data given here on metals in Israeli canned food, Australian fish, Polish margarine and Zambian gin show the international nature of the problem. Water supplies in Boston as well as in Glasgow, as will be seen, can carry the same contaminating metals as do illicit spirits in Kentucky. For this reason summaries of food laws relating to metals from the UK, the US, Australia and other English-speaking countries, as well as the current international codes, are given in the text. It is hoped that, in this way, the basic data on metals in food and their significance for human health will be related to the practical context of the day-to-day user and the food producer.

Gratitude is owed to many for help given in writing this work: to my wife Ann, especially, for encouragement and practical help with references and bibliography; to Drs George and Olga Berg of the University of Rochester, New York, for setting me right on US food legislation, and for other advice and help; to my brother Brian who, from the point of view of his own profession of engineering, contributed to sections on metal contamination during food processing; and to many others who will remain anonymous. Finally, a word of gratitude to the hard-working secretaries who deciphered my writing and endured my re-editing with so much patience.

CONOR REILLY
Brisbane, 4 March 1980

CONTENTS

CONTENTS

PART I

General

CHAPTER 1

THE METALS WE CONSUME

'Food and nutriment are not synonymous; there is much in food that does not nourish the body'.

McLAREN[1]

The small amount of grey-white ash that remains in the bottom of a crucible after a food sample has been incinerated in a muffle furnace is a familiar sight to all who have followed even an elementary course in food science. This is the 'ash' listed in most standard food composition tables. For more than 150 years, from the time when techniques for the proximate analysis of food were developed by scientists at the Weende Agriculture Institute in Germany, 'ash' has been the global term used to record the inorganic components of foods. What those components are, and what their significance might be for consumers, have only relatively recently been given detailed attention. We now know that their significance can be very great indeed, even vital to life.

Two other German scientists, the physiologists Voit and Rubner, whose views dominated the emerging science of nutrition at the beginning of this century, stressed the key roles of carbohydrates, proteins and fats, for the maintenance of health. They did not totally ignore that other component of food, the ash, which their colleagues at Weende determined so meticulously but, as has been observed by a modern commentator, for them, ash had little if any significance[2]. This view was shared by many other scientists of their time and persisted into the twentieth century. Even as late as 1910, the United States Department of Agriculture officially maintained that carbohydrate, fat and protein were the only nutrients essential in the human diet[3].

Many decades were to pass from the time when scientists first began to

1

analyse food for its components, before the role of vitamins as essential for prevention of a variety of diseases was to be established. This was achieved in spite of the stubbornly held belief that most illnesses, including several now clearly recognised as diet-related, were caused by infection. No less difficulty was to be met in establishing the equally essential role of 'ash'.

The importance of certain inorganic elements for the body's vital functions had, in fact, been accepted by some scientists for some time. Iron had been identified in body tissues as far back as the early eighteenth century. It was just over 100 years later, in 1820, that the association of iodine with goitre was recognised[3]. Almost another century passed before the roles in animal health of a number of metals, such as cobalt and molybdenum, were established. That these elements played a similar role in humans, took much longer to prove. Zinc, for instance, has only been accepted as an essential nutrient since the early 1960s. The essentiality of other inorganic elements has been established subsequently through a combination of epidemiological observations, clinical trials and biochemical investigations. There is still uncertainty, however, as to whether a few elements such as vanadium and tin should be included among human nutrients.

As McLaren noted, not everything in food is nutritious. Indeed, some of its inorganic components are capable of causing serious damage to health. Several of these elements occur naturally in food; others are adventitious. These are the contaminants, part of the 'undesirable materials which have been added inadvertently before, during, or after processing of food'[4]. Contaminants have in the past and continue today to cause considerable concern to health authorities, manufacturers and suppliers of food and, not least, to its consumers.

The term 'contaminant' has different meanings for different people. For the food producer, it can imply such impurities as arsenic, lead and other heavy metals whose concentrations must not exceed levels which comply with legal standards or, in their absence, with accepted codes of practice. The consumer thinks of contamination as the addition to his food of anything that would make it unfit or injurious to health, and not of the quality he has the right to expect. The food producer cannot merely think of the legal aspects of contamination. He must also pay heed to the views and the expectations of the customer.

In this age of instant information and insistent media intrusion into our lives, people know far more, though they are not necessarily better informed than were their predecessors, about food and what it may contain. Words like dioxin and nitrosamines, and such concepts as the greenhouse effect and acid rain production, known only to a few experts not many years ago, are now part of the vocabulary of the general public. Familiarity with these less desirable aspects of technological progress has bred, not as traditional wisdom would have it, contempt, but concern and even, for some, fear. Contributing in no small measure to that fear are the often alarmist reports in the written and electronic media on toxic contaminants in food.

The list of poisonous components of food about which the public is warned, is long. It includes many organic as well as inorganic substances, and usually prominent among the latter are the metals.

Not many years ago Anthony Tucker in his popular book *The Toxic Metals*, warned that 'heavy metals are among the most dangerous and least understood of contaminants. Because they exist naturally as part of the earth's crust they occur in all soils, rivers and oceans. In the right quantities some are essential to life. Others are so poisonous that only a few millionths of a gram can kill. Many that are capable of disrupting living processes are in widespread industrial use and, as contaminants, are extending through the biosphere, so that in increasing quantities they distort the naturally occurring distribution of metals to form an accidental but potentially disastrous addition to the diet of all living things'[5]. Tucker expanded on these chilling words. The titles of his chapters indicate his approach: *Mercury strikes again — the sickness of Minamata spreads to Scandinavia, Canada and the United States*, and, *The metals you need, and the metals you do not want at any price — even if the economists say they are good for you*. The book finishes with an even more depressing chapter: *How public health authorities turn a blind eye yet pretend to protect you. The road to disaster*.

Tucker's views have been repeated many times in popular articles and books. They are accepted by many people who, possibly as a result of the fears generated by such reading, advocate the abandonment of all processed and factory-farm food, and a return to a diet which is natural, whole and unprocessed, and consequently one which they believe to be 'uncontaminated'. They would strongly defend themselves against the criticism made by the American toxicologist, Liener, that their expectations of pure food are 'unreasonable'[6].

It is not just the popularisers who warn of the menaces of the metals in our diet. Professional scientists, writing in far from sensationalising journals, can do the same. Jerome Nriagu, of the Canadian National Water Research Institute, issued dire warnings recently in the reputable journal *Environmental Pollution*[7]. His title asked the question: *A Silent Epidemic of Environmental Metal Poisoning*? His report advanced 'the provocative view that the current levels of toxic metals in some environmental compartments may be high enough to constitute a threat to human health. Some may argue that such a proposition cannot be substantiated by hard scientific data. Nevertheless, it is incontestable that each environmental compartment has a limited carrying capacity for metal pollution and, with enough time, the current rates of metal inputs will become stressful to many ecosystems'. Nriagu believes that several million people worldwide suffer currently from subclinical metal poisoning and that a major cause of this is metal contamination of foods and beverages.

This is indeed, a provocative view and undoubtedly could serve to promote and stimulate further discussion as, the author claimed, was the purpose of his report. However, there are also less alarmist views, usually representing the official attitude of governments and health authorities. F.M. Strong, Chairman of the Sub-Committee on Naturally-occurring Toxicants in Food of the US National Research Council[8], had the following to say: 'People need to understand that there is no reason to fear a particular food chemical simply because an overwhelming excess of it may be harmful. Ingestion of chemicals that would be toxic if consumed in excessive amounts is a perfectly normal situation that mankind has always faced and inevitably will'.

In practice many consumers accept Strong's view as reasonable. However, it is a view that does not absolve health authorities and those responsible for food safety from constant vigilance. Our ever increasing analytical abilities and understanding of toxicology will not allow us to be complacent about the safety of those adventitious, inorganic components we find in our food. There may not be, in reality, a 'silent epidemic of environmental metal poisoning'[7], but there is an enduring need to be alert to the potential dangers to public health of the metals that get into our food.

The American toxicologist, Frederick Oehme, in a study of the toxicity of heavy metals in the environment, has drawn attention to two characteristics of the metals; their strong attraction to biological tissues and their slow elimination from the body, which make them particularly hazardous for human health. He noted that 'it is becoming increasingly apparent, due to more sophisticated methods of clinical and biochemical evaluation, that low environmental levels of heavy metals may produce subtle and chronic effects previously unrecognised.... The casualness with which environmental pollutants may produce their effects raises the problem of identifying these changes, characterising them for clinical diagnoses, developing treatment procedures and, finally, disseminating this information as promptly as possible'[9].

It is perhaps appropriate to note here that not all contaminating and adventitious metals are unwelcome in the diet. Several can make a significant contribution to good nutrition. To some extent we have become dependent on contamination to meet our needs for several of the inorganic trace nutrients. A highly refined, pure diet can lack certain essential nutrients which in other circumstances might be present as adventitious components of the diet. There is good evidence that our intake of iron, zinc and chromium may be boosted effectively by contributions from cast and galvanised iron and stainless steel used in the manufacture of utensils with which our food comes into contact[10]. Nevertheless, such adventitious pick-up of metals by food can also be excessive. In some parts of Africa, the use of iron cooking pots helps to prevent anaemia but can also lead to iron overload of the liver. Consequently, there is need for balance between what is a safe and an excessive intake of metals from whatever source they come[11].

It is for such reasons that most countries have specialist committees with ongoing briefs to monitor the status of metal contamination of foods and to recommend new or amended legislation whenever this appears called for. The tasks that such committees face are, as this volume will show, considerable, for in spite of all our progress, the question of metals in food and their relation to human health is complex and remains far from a final resolution.

THE METALS

Some 80 of the 104 elements now listed in the Periodic Table are metals. Another 17 are described as non-metals, leaving a small intermediate group called metalloids. The distinctive physical and chemical properties of the metals and non-metals are summarised in Table 1[12]. The differences in chemical properties between the metals and non-metals are mainly related to the fact that atoms of non-metals can readily fill their valence shells by sharing electrons with, or transferring electrons to, other atoms. Metallic characteristics decrease and non-metallic characteristics increase with the number of valence electrons. In addition, metallic characteristics increase with the number of electron shells. Thus the properties of the

TABLE 1
Properties of Metals and Non-Metals

Metals	Non-metals
Physical properties	
1. Good conductors of heat and electricity	1. Poor conductors
2. Malleable and ductile in solid state	2. Brittle, non-ductile in solid state
3. Metallic lustre	3. No metallic lustre
4. Opaque	4. Transparent or translucent
5. High density	5. Low density
6. Solids (except mercury)	6. Gases, liquids or solids
7. Crystal structure in which each atom is surrounded by 8–12 near neighbours; metallic bonds between atoms	7. Molecules consist of atoms covalently bonded; the noble gases are monatomic
Chemical properties	
1. One to four electrons in outer shell; usually not more than three	1. Usually 4–8 electrons in outer shell
2. Low ionisation potentials; readily form cations by losing electrons	2. High electron affinities; readily form anions by gaining electrons (except noble gases)
3. Good reducing agents	3. Good oxidising agents (except noble gases)
4. Hydroxides basic or amphoteric	4. Hydroxides acidic
5. Electropositive; oxidation states positive	5. Electronegative; oxidation states either positive or negative

succeeding elements change gradually in progressing across the Periodic Table, as the number of valence electrons change from left to right, and down the groups, when the number of electron shells increases. There is not, therefore, a sharp distinction but rather a merging of characteristics, between the metals and the non-metals. The result is that there is an intermediate group of elements which show characteristics both of metals and non-metals. These are the metalloids, a group which includes boron, silicon, germanium, arsenic, selenium, antimony and tellurium. Because of their metallic characteristics, and their importance as contaminants in

TABLE 2
Average Abundance of Some Metals in the Earth's Crust

Metal	Lithosphere content (mg/kg dry wt)	Soil content (mg/kg)
Iron	50 000	7 000–55 000
Manganese	1 000	200–5 000
Chromium	200	5–3 000
Vanadium	150	20–25
Nickel	100	10–800
Zinc	80	10–300
Copper	70	2–100
Cobalt	40	1–5
Molybdenum	2	> 1–2

food, some of these will be included in this study, without any particular distinction being made between them and the true metals.

METALS IN THE ENVIRONMENT

Analysis of the tissues of the human body will show the presence of most of the metallic elements, some in reasonably large amounts, others just detectable as traces. This is not surprising since the food we consume also contains a similar wide range of metals, reflecting their distribution in the environment. The soils in which plants grow contain metals from a number of sources, principally, the rocks from which the soil was formed, as well as from fertilisers, waste and other materials added by man. Metals are contributed to soil by the debris of mining and industrial activities, by the dust and smoke of fossil fuel combustion, and by other forms of pollution. Water, too, makes its contribution, to an extent related to the source of supply and the degree of pollution.

The actual amount of metal found in any soil sample will depend primarily on the nature of the parent rocks from which it was formed. Table 2 shows the estimated overall abundance in the lithosphere of a number of metals and the range of concentrations found in soils. The distribution in soils of some of the less common, though nevertheless widely distributed, metals is shown in Table 3.

TABLE 3
Average Trace Metal Content of Soil

Metal	Soil content (mg/kg)
Antimony	6
Arsenic	6
Barium	500
Beryllium	6
Cadmium	0·06
Lead	10
Mercury	0·3
Scandium	7
Selenium	0·2
Silver	0·1
Tellurium	10·1
Tin	10
Vanadium	100

The atmospheric distribution of metals and other elements often reflects industrial activity rather than geological structures. The world was made acutely conscious of this in April 1986, when the nuclear disaster at the Chernobyl reactors, near Kiev in the USSR, resulted in contamination in parts of Europe as far distant as the sheep farms of Wales, and the tomato fields of Italy.

Non-radioactive metals can also be widely distributed from their point of origin. A detailed study[13] of levels of 30 elements in the atmosphere in the UK, showed how even distant industrial activity could contribute to the metal content of the air. While levels of lead, mercury and zinc were 380, 0.09 and 415 ng/kg respectively in the industrial city of Nottingham, they were still appreciable at 139, 0.13 and 132 ng/kg at rural Lake Windemere. The atmosphere at Windemere, in its turn, had levels far exceeding those reported for rural sites in Canada.

Concentrations of metals in water may reflect nearby industrial activity as well as the composition of local rock and soil. In addition, reticulated water may carry metallic contributions due to the composition and the condition of pipes and storage tanks. Table 4[14] shows the levels of some metals in domestic water supplies in a city in Australia. While these are the levels which are expected in well managed municipal supplies, and which conform to WHO standards[15], actual levels in individual homes can be considerably higher, due to the types and condition of the plumbing and fittings. A recent US study[16] found that brass valves and fittings were

TABLE 4
Metals in Domestic Water (mg/litre) (Brisbane, Australia)

Metals	Mains water	Cold tap	WHO standards
Chromium	< 0·02	< 0·02	0·05
Nickel	< 0·01	< 0·01	—
Copper	0·044	0·05	0·05
Zinc	< 0·01	0·21	5·0
Iron	0·15	0·55	0·1
Manganese	0·07	0·13	0·05
Cadmium	< 0·002	< 0·002	0·01

responsible for widescale lead, copper and zinc contamination of domestic water supplies.

METALS IN FOOD

The metal content of food, whether it be of plant or of animal origin, will depend on many factors, ranging from environmental conditions to methods of production and processing. Even in the same class of food there can be considerable variations in concentrations of metals. Zinc in oysters, for instance, may range from less than 1 to more than 1000 mg/kg, and nickel in meat from zero to 4.5 mg/kg[17]. Even within a single sample of a food, metals may not be evenly distributed so that there can be differences between levels in different parts. For example, while the germ of a grain of wheat contains 7.4 mg/kg of copper, its endosperm may have less than 2.0 mg/kg. It is for this reason that refining or fractionation of cereals that occurs during milling affects the concentrations of metals in the food as consumed. While unpolished rice contains 0.16 mg/kg of chromium, the polished product has only 0.04 mg/kg. In the same way when wheat is converted into flour, its iron is reduced by almost 76 per cent, manganese by 86 per cent and cobalt by 89 per cent[18].

Different brands of the same kind of processed food can have considerably different levels of metals. A Canadian study showed that the lead content of different brands of infant formulae ranged from 1.1 to 122 μg/kg, while for cadmium it was 0.05 to 7.55 μg/kg[19]. Levels of chromium in canned fruit and vegetables sold in Sweden ranged from less than 0.005 to 0.43 mg/kg. The top level found in comparable fresh foods was 0.032 mg/kg[20].

The place of origin of a food can have an influence on its metal content.

This is, of course, a reflection of environmental variations, particularly with relation to soil composition. Such geographical factors can have considerable, and sometimes unexpected, nutritional significance, as has been seen in the case of selenium. Health authorities in Finland became concerned in the early 1970s when surveys revealed a very low average intake of this essential trace element in the diet[21]. This was found to be due to the low level of selenium in Finnish soils and, consequently, in cereals and other locally produced food[22]. However, there were inconsistencies in the findings. In some years, levels of selenium in the diet rose significantly. This was found to occur in years when the local cereal crop was poor and it was necessary to import wheat from overseas. The source of this wheat was the US where soil selenium levels in the growing areas are high. In an enterprising move, Finnish authorities, recognising the connection between uptake of selenium by crops and the level of the nutrient in the diet, decided to bring Finnish levels up to those found in the US imports by adding the mineral to fertilisers used on farmland. This has now been done for 10 years. The result is that average selenium dietary intake levels in 1986 was three to four times higher than in the mid 1970s[23].

It is interesting to note that the reverse situation has, apparently, occurred in the UK. A recent study has found that dietary levels and blood concentrations of the element have decreased in parallel with a reduction in the use of imported US and Canadian grain, and their replacement by European-grown stocks. The local product has a selenium content 10 times lower than that of the American wheat[24]. This is an unexpected outcome of the success of farming expansion and promotion of agricultural self-sufficiency in the EEC.

HOW MUCH METALS DO WE ACTUALLY EAT WITH OUR FOOD?

Knowing the mean and the range of concentrations of various metals in foods does not immediately tell us a great deal about the amount of a particular metal that is actually consumed by a population group, and still less by an individual. Further data must be gathered and calculations made before we can assess these intakes with an acceptable degree of certainty. This is a fact that certain journalists and some consumer activists apparently fail to appreciate. The findings, for instance, that cayenne pepper can contain as much as 30 mg/kg of aluminium[25] and that some home-grown vegetables have 0.1 mg/kg of thallium in their edible portions[26] do not mean that those of us with a taste for spicy food are in danger of developing Alzheimer's disease from excessive intake of aluminium, or that the home vegetable patch is implicated in thallotoxicosis. The data do not justify a sensational headline in the press nor a protest to the Department of Consumer Affairs.

It should be recognised that though such levels of metals could, under unusual and rare circumstances, pose a health threat to some people, in general none of us would normally consume sufficient spices or vegetables grown on thallium-rich soil to cause a problem. But this is not to say that a knowledge of average as well as extreme levels of metals in foods is not important in assessing the public health consequences of food contamination. These are the data which, in fact, health authorities and other regulatory bodies use when they are assessing the risk factors resulting from heavy metals and other residues that occur in foods. A similar approach is used when deciding on legal standards for metal-containing and other permitted additives in food.

Methods for estimating the dietary intake of chemicals in food and, in particular, of calculating their PDI (or potential daily intake) are described in detail by Lindsay[27]. The points he makes which are particularly pertinent to metals in food are summarised here.

The actual level of exposure of a consumer to a potentially toxic substance, such as a heavy metal, is not directly related to the maximum, or even the legally permitted level of the substance in foods consumed. This can only be determined if, in addition to information on concentrations of the metal in foods, data are gathered on the amounts of metal-containing food consumed by individuals.

Food consumption data are generally obtained either by studying the food consumption patterns of a population group as a whole, or of selected individuals within a population. Several different methods are used to do this.

Food diaries, weighed intake studies:
Persons being studied are asked to keep a diary of the type and amounts of foods consumed over a 24-h or longer period. Amounts are estimated or, if possible, weighed. If the sample is large and representative of the population, this method gives reasonable data on the long-term average consumption of foods within the population group. If the study is performed on a small group, or if data on individuals are sought, a duplicate portion study is particularly valuable. In this method, the individual keeping the record buys and prepares twice the usual portions

of food consumed. The duplicate is weighed and stored for subsequent analysis.

Dietary recall:
Participants are asked to recall the types and amounts of food consumed, usually over the previous 24-h period. This is less accurate, but requires less effort and usually has a good response rate compared to the food diary approach.

Food frequency:
This obtains data on the usual patterns of consumption for individual types of food. It is semi-quantitative in nature but useful for collecting retrospective data on certain food intakes in epidemiological studies.

Food disappearances methods:
Either at the household or the national level, these give a broader estimate of food consumption by the general population. In the former, the amount of food that 'disappears' out of the home larder is recorded, usually once a week. This is divided by the number of members of the household, with corrections made for meals eaten away from home as well as for food wastage. The national estimate is based on official records of food production, with imports added, less exports, food used for other than human consumption and wastage. Apparent individual consumption is found by dividing the quantity of food consumed by the total population.

The total diet (or market basket) study:
This is a somewhat different approach, which achieves the same general purpose as the above methods. The method has been widely used in the UK, the US and several other countries to obtain estimates of the average intakes of essential food constituents and contaminants. A 'market basket' of food, reflecting the food consumption patterns of the general population, is purchased in normal retail outlets, prepared for consumption by ordinary cooking methods, and analysed either individually or combined into one or more food group composites (such as cereals, meat, root and other vegetables, etc.) in proportions based on available consumption data. The method is particularly valuable in that analyses of a relatively small number of samples allows an approximate daily intake to be calculated. Analysis of every individual food would place an unacceptable demand on laboratory resources. These methods provide

information on average or 'per capita' levels of intake of food but fail to address sections of the population whose food consumption patterns diverge widely from the average. These are the groups or individuals who consume extreme amounts of a particular type of food or foods in general and those, like infants, who consume a restricted range of foods. Such people may, in fact, be especially at risk from a metal or other contaminant in a foodstuff.

While surveys on individuals can provide data for extreme consumers, it is usually not possible to survey a sufficiently large sample from which to select data representing the consumption patterns of 'extreme' habits or habits atypical of normal patterns. The difficulty can be overcome by using data which are hypothetical but based on previous, well-conducted food consumption and energy intake studies.

Experience shows that the total energy in all human diets is self-limiting. Data published by the World Health Organization[28] show that, for example, the energy requirement for an adult 80-kg man who is exceptionally active, is always approximately 1.6 times the energy requirement of a moderately active 65-kg man. Moreover, the 95th percentile energy intake of a particular age–sex class never exceeds twice the average intake of the same class of people[29]. As Lindsay has noted in his paper, 'while those energy needs could be made up from a considerable variation in intake of individual foods, it is reasonable to assume that the range of the total quantity of food consumed by individuals in a population varies by no greater than a 2-fold factor. Thus the data available on the total average food consumption patterns can be used to calculate the likely extreme total intake of food in that population through application of a 2-fold factor'.

Food consumption data obtained by the methods described above provide the basis for the calculation of PDIs of the various metal contaminants in the diet.

A good example of the use of some of the above methods to estimate intake of a metal in the diet is given by Smart and Sherlock in their study of chromium in the UK diet[30]. They used results from the UK total diet survey. The foods were combined into nine groups and then analysed for chromium. A similar use of a total diet or market basket survey to estimate zinc in the Australian diet was reported by English[31].

The need for care when comparing intake estimates which have been derived by different methods must not be overlooked. This was stressed by Louekari and his colleagues in their study of cadmium intake by a group of Finnish men[32]. They used two of the methods described here,

TABLE 5
Trace Elements in Body Fluids and Tissues

Metal	Diet (mg/day)	Urine (mg/day)	Sweat (mg/day)	Hair (µg/g)	Milk (µg/litre)
Essential					
Chromium	0·025–0·050	0·008	0·059	0·69–0·96	10–29
Manganese	2·0–9·0	0·225	0·097	1·0	7–120
Iron	8·0–20·0	0·25	0·5	130	150–1500
Cobalt	0·15–0·60	0·26	0·017	0·17–0·28	1–8·6
Copper	1·2–3·2	0·06	1·59	16–56	197–500
Zinc	13·0–25·0	0·5	5·08	167–172	316–5300
Selenium	0·03–0·10	0·04	0·34	0·3–13	13–21
Molybdenum	0·12–0·30	0·15	0·061	—	—
Nickel	0·15–0·2	0·011	0·083		20–83
Possibly essential					
Vanadium	0·01–0·03	0·015	—	—	—
Toxic					
Cadmium	0·010–0·080	0·03	—	—	2–19
Lead	0·070–0·300	0·03	0·256	2·8–4·8	12–30
Mercury	0·002–0·030	0·015	0·0009	18–19	1–6
Antimony	0·25–1·50	0·07	0·011	6	—
Beryllium	0·012–0·100	0·0013	—	6·5	—
Arsenic	0·055–0·090	0·195	—	—	36
Barium	0·01–1·25	0·023	0·085	2	20
Non-toxic					
Tin	2·92–3·08	0·023	2·23	—	20
Aluminium	9·0–14·0	0·1	6·24	5	<5–45
Titanium	0·003–0·60	0·33	0·001	0·05	20
Zirconium	3·5–4·0	0·14	—	—	—
Boron	2·0–20·0	1·0	—	7	—

calculations based on food composition tables stored in a computer data base, and direct analyses of duplicate diet samples. Intake estimated from the data base was 15.8 μg/day and from direct analyses of duplicate portions 8.2 μg/day. The authors stressed that several complicating factors can affect the reliability of both estimates. It is clear that we are still a long way from being able to obtain absolutely reliable answers to our questions about intakes of trace elements by all sections of the community.

THE METAL CONTENT OF THE HUMAN BODY

It is not surprising that since the environment in which we live, the air we breathe, the water we drink and the food we eat contain a wide variety and range of concentrations of metals, that our bodies should also display a range in kind and content of these same inorganic elements. Most of the metals in our bodies has come in with food. However, not all the metal we consume is retained. Some passes straight through and is lost in faeces. Some more is first absorbed and then lost in sweat, urine, through bilary excretion, in discarded hair and skin. The amounts of metal actually absorbed by the body from food in the gastrointestinal tract will depend to some extent on our choice of diet, but also on our state of health and our genetic make up.

It will be affected too by the presence of other factors in our diet. Some naturally-occurring components of food, such as phytates and fibre can inhibit absorption; others such as ascorbic acid, a variety of other organic compounds, chelating agents and acids, can help. There can also be competition for absorption between metals — copper with iron and zinc, selenium with mercury, molybdenum with copper. In spite of such a variety of influences on metal uptake, it is possible to draw up a table of average quantities indicating dietary intake levels in the whole body as well as in several tissues, and amount lost in fluids, which gives a very approximate, though still useful picture of the status of metals in human nutrition, as is shown in Table 5. Though the data have been gathered from a wide variety of sources it is seen that there are still many gaps in the table, due especially to technical difficulties in analysing biological tissues for very low levels of several elements. These difficulties will be discussed in Chapter 4.

Within the body itself the distribution of metals is by no means uniform. Some metals are accumulated in particular tissues and organs and others

TABLE 6
Classification of Metals in the Body

Grouping	Metal	$g/70\ kg^a$ (reference man)
1. Macronutrient metals essential for body function	Calcium	1 700
	Potassium	250
	Sodium	70
	Magnesium	42
2. Micronutrient metals essential for body function	Iron	5
	Zinc	3
	Manganese	< 1
	Copper	< 1
	Molybdenum	< 1
	Cobalt	1
	Chromium	< 0·1
	Silicon	1
	Nickel	< 0·1
	Selenium	< 0·01
3. Metals for which essentiality has not yet been established, although there is evidence of their involvement in some cell reactions	Barium	< 1
	Arsenic	< 1
	Strontium	< 1
	Cadmium	< 1
	Vanadium	< 1
	Tin	< 1
4. Metals found in the body for which no metabolic function is known	Lead	< 1
	Mercury	< 1
	Gold	< 1
	Silver	< 1
	Bismuth	< 1
	Antimony	< 1
	Boron	< 1
	Beryllium	< 1
	Lithium	< 1
	Gallium	< 1
	Titanium and others	< 1

[a] Based on Williams, D.R. (1983). *J. Intern. Metab. Dis.*, 6 Suppl. 1, 1–4.

in different portions of the body. For example, the kidney will preferentially accumulate cadmium, but not lead, which builds up in blood. Chromium is concentrated preferentially in muscle tissue, with only a minor fraction going to the kidney.

THE ROLE OF METALS IN THE HUMAN BODY

Among the many metals found in the human body, only a small number are believed to be essential for normal life. Deficiency of any one of these essential metal nutrients will result in specific biochemical lesions within cells of the body and the development of characteristic clinical symptoms. The symptoms will normally respond when the deficiency is corrected by supplying an adequate amount of the missing element. Most of the other metals present are artefacts, with no nutritional significance. In some cases these other metals have the potential for deleterious effects on body function.

The nutrient metals are traditionally divided into two classes according to the amounts of each which are required for normal function. As is shown in Table 6, potassium, magnesium, calcium and sodium are classified as macronutrients and the remainder as micronutrients.

The use of the term 'trace element' for this second group is no longer as common as it was in the past. As has been noted by Versieck and Cornelis[33], the term originated in the days when the minute amounts of some elements in biological tissues, micrograms to picograms per gram wet weight, were beyond the instrumental capabilities of analysts to quantify exactly. Consequently, they were referred to as occurring in 'traces'. Today the term is usually used for an element occurring at a level of <0.01 per cent (<100 μg/g) Other terms are used by some workers as a replacement for trace element, such as minor element, oligoelement and, as here, micronutrient.

The term 'ultratrace' element is also met with occasionally as a description of metals that occur at very low levels. It is usually, though not consistently, used to describe elements that occur at levels <0.01 μg/g (<10 ng/g).

Use of the term 'trace element' can, in fact, cause some confusion. Some of the trace elements so-called are actually present in food and the body in relatively significant amounts. It is the dietary requirements for them that are small. Apart from the question of quantity, as Versieck and Cornelis have noted, for a long time the term also held a connotation of uncertainty as to the biological significance of the element. Probably for this reason, iron, even though it occurs in the body at a level of <0.01 per cent, is often considered to be distinct from the group of trace elements.

An examination of the literature on nutrition published over the past few decades shows how views on the second category of metals have changed. Up to the mid 1950s only six of these elements were identified as essential: iron, copper, zinc, cobalt, manganese and molybdenum. Since then, as investigations proceeded and better methods of analysis were developed, several more metals have been recognised as essential, as is seen in the table.

There is no reason to suppose that the third group of metals is definitive. It would not be surprising if some elements with currently unknown physiological functions were also found to have specific biochemical functions in the body. There is growing evidence that arsenic, for instance, is one such element.

There are several reasons why there is still uncertainty regarding the role of metals in the body. While very important, analytical problems are not the only reason for this situation. Development of suitable experimental protocols for investigating the essentiality or otherwise of a metal is particularly difficult. It is relatively easy to demonstrate that a dietary insufficiency of a macronutrient will result in metabolic disorders, but it is not so for a micronutrient. For ethical reasons we cannot subject humans experimentally to deprivation of certain elements in the diet in order to see what symptoms will result. Quite properly, ethical committees frown upon the once widely practised use of volunteers, even of the researchers themselves, for such investigations. Animal subjects do not always adequately match human requirements and responses to the element being tested.

Experimental and analytical difficulties mean that for a long time to come the exact composition of the second and third groups of the elements in the body will remain uncertain. Even the fourth group, metals found in the body with no known beneficial metabolic function, may change in the same way. This is what happened to selenium, whose notorious toxicity was well known before new evidence showed that it should be moved into the essential micronutrient group. It has been suggested that cadmium should undergo the same re-classification[34].

To some extent the known functions of inorganic macronutrients and micronutrients overlap. Both groups of nutrients work in three main ways: as constituents of bone and teeth; as soluble salts which help to control the composition of body fluids and cells; as essential adjuncts to certain enzymes and other functional proteins. The macronutrients play major roles in the first two functions, while the micronutrients are especially prominent in assisting enzyme function. Very few of the proteins that act as biological catalysts can do so entirely on their own. Most need the assistance of a non-protein prosthetic group. If the prosthetic group is detachable it is known as a coenzyme. The group may be an organic molecule, containing one or more atoms of a trace metal, or it may consist solely of a trace metal. In the latter case, if the metal is detachable from the protein part, it is known as an activator.

Iron and copper occur in the prosthetic group of many enzymes concerned with cellular oxidation–reduction reactions. Molybdenum has much the same function in other enzymes concerned with oxidation. Zinc and manganese function as detachable activators on some enzymes and as part of the prosthetic group in others. Most of the other inorganic micronutrients are believed to play similar enzymatic roles and, as a consequence, the enzymes in question are often known as metalloenzymes.

Trace elements are also found in other important functional components of the body, including hormones and vitamins. The production and storage of insulin in the beta cells of the pancreas, for instance, involve the metal zinc. Haemoglobin and myoglobin, oxygen transporters of blood and muscle tissue, are iron-containing compounds. Cobalamin or vitamin B_{12} contains cobalt. We shall look at some of these compounds and the related functions of the metals in later sections. For the present it is sufficient to indicate briefly that those metals which are included in groups 1, 2 and 3 of Table 6 can be shown to play essential roles in human metabolism and body function. This cannot be said by any means of all the metals which occur in food.

THE TOXIC METALS

As will be made clear later, it is not always possible to draw a distinction between essential and toxic metals. All metals are probably toxic if ingested in sufficient amounts. Indeed, sometimes the margin between toxicity and deficiency is very small, as is the case with selenium. In addition, it is difficult to consider toxicity of a single metal in isolation. Under normal conditions all metals are capable of interacting in the body if consumed. The physiological effects, including its toxicity, of cadmium, for example, are closely related to the amount of zinc also present. Likewise the function of iron in cells is affected by both copper and cobalt and, to some extent, also by molybdenum and zinc. Several other similar interactions between metals in the body are known, suggesting that caution must be exercised when deciding whether or not a particular trace metal is toxic.

In spite of the above reservations it is possible to differentiate between elements that are known with certainty to be essential and those which display severe toxicological symptoms at extremely low levels and have no known beneficial functions. Mercury, cadmium and lead are usually considered to qualify for inclusion in the latter group, though, as has already been noted, it has been suggested that cadmium might one day be listed as an essential element. These three are among the most commonly encountered toxic metals in food. All three have on occasion been responsible for spectacular and devastating outbreaks of poisoning.

The term 'heavy metal' is often used to describe the group of toxic metals which includes mercury, lead and cadmium. It is a descriptive rather than an accurate scientific classification. It is usually taken to refer to metals of a high specific gravity which have a strong attraction to biological tissues and are slowly eliminated from biological systems[35]. However, a typical listing of heavy metals will often be found to include arsenic, beryllium, boron, selenium and other metals and metalloids, as well as mercury, cadmium and lead. This is probably a convenience and relates more to their toxicity than to specific gravity.

All of these elements, whether true 'heavy metals' or not, are considered to be toxic, or at least to have the potential to become so under certain conditions. It is useful here to consider what we mean by the term 'toxic'. Phipps[36] defined a toxic metal as one which belongs to a 'group of elements which have neither an essential nor a beneficial but a positively catastrophic effect on normal metabolic functions, even when present in only small amounts'. What exactly these catastrophic effects are in individual cases is not by any means always clear. Phipps noted that the pathological effects and significant dose in metal poisoning are remarkably complex and variable, depending particularly on factors that can modify the uptake of the metal as well as those which control its subsequent metabolism. We shall see how variable the effects of toxic metals can be and how great is the difficulty of deciding when toxicity actually begins.

EFFECTS OF METALS ON FOODS

The concern of the food processor is not merely to see that his products are free of toxic metals or even of essential metals in quantities high enough to cause poisoning; nor is it simply to meet the various legal requirements or codes of practice that apply to metals in food. He has also to make sure that his products do not contain metals which might bring about deterioration and quality defects. He must, moreover, be alert to the possibility of pick-up from non-food sources such as containers and plumbing, of traces of metal that would affect the colour, flavour or shelf-life of his foodstuffs: for trace metal contamination can result in any or all of these effects.

Trace metals can cause a variety of colour changes in food during cooking and storage. Even at levels of only a few milligrams per kilogram, complexes can be formed between metal ions and organic compounds causing the development of colours, not all of which are welcome. There was a time when less reputable cooks and food processors used the complex formed between copper and chlorophyll to give leafy vegetables a bright and stable green colour — either by cooking in unlined copper saucepans or even by dropping a copper coin into the pot! Copper will also, at very low concentrations, darken cherries and even turn them black by forming another organo-metal complex. Traces of iron can also be responsible for colour changes in food. The metal reacts with anthocyanins in some fruits to produce a black pigment. It can impart a grey-green colour to cream and have a similar effect on chocolate-containing foods. Aluminium and tin can also cause darkening of food colours.

The flavour of a food product can also be adversely affected by the presence of a metal contaminant, picked up, for example, from can corrosion. This was shown by Borocz-Szabo[37] in a carefully-designed study of the influence of iron on the sensory qualities of a variety of liquid foods, including fruit juices, milk, beer and wine. Corrosion products of steel, especially iron salts, caused loss of odour as well as development of astringent, metallic or bitter tastes. Milk, in particular, developed a nauseous taste under the conditions investigated.

Another effect of trace metal contamination of food, and which causes considerable losses at times in food processing, is the production of rancidity in fats. Traces of copper, iron and some other metals can act as catalysts in the oxidation of double bonds in the unsaturated fatty acids of oils and fats. The result is rapid and costly deterioration of cooking oil and fat-containing foods.

REFERENCES

1. MCLAREN, D.S. (1981). *Nutrition and its Disorders*, 3rd edn (Churchill-Livingstone, Edinburgh).
2. GRIGGS, B. (1988). *The Food Factor*, 22 (Penguin, London).
3. GUTHRIE, H.A. (1975). *Introductory Nutrition*, 4 (Mosby, St Louis, Mo).
4. PEARSON, D. (1976). *The Chemical Analysis of Food*, 7th edn, 684 (Churchill-Livingstone, Edinburgh).

5. TUCKER, A. (1972). *The Toxic Metals*, 9 (Pan/Ballantine, London).
6. LIENER, I.E. (1969). *Toxic Constituents of Plant Foodstuffs* (Academic Press, New York and London).
7. NRIAGU, J.O. (1988). A silent epidemic of environmental metal poisoning? *Environ. Pollut.*, 50, 139–61.
8. COMMITTEE ON FOOD PROTECTION (1973) *Toxicants Occurring Naturally in Foods*, 2,3 (National Academy of Sciences, Washington, DC).
9. OEHME, F.W. (1978). *Toxicity of Heavy Metals in the Environment* (M. Dekker, New York).
10. REILLY, C. (1987). Can food be too pure? The case of trace metal contamination. *Food Tech. Aust.*, 39, 27–8.
11. REILLY, C. (1985). The dietary significance of adventitious metals in food. *Food Add. Contam.*, 2, 209–15.
12. NEBERGALL, W.H., SCHMIDT, F.C. and HOLTZCLAW, H.F. (1968) *General Chemistry*, 3rd edn, 530 (Raytheon Educ. Co., Boston, Mass.).
13. PEIRSON, D.H., SALMON, L. and Cambray, R.S. (1973) *Nature* (*London*), 241, 252–6.
14. CRAUN, G.F. and MCCABE, L.J. (1975). Problems associated with metals in drinking water. *J. Am. Water Works Assoc.*, 67, 593–9.
15. *WHO INTERNATIONAL STANDARDS FOR DRINKING WATER* (1971). (WHO, Geneva).
16. SCHOCK, M.R. and NEFF, C.H. (1988). Trace metal contamination from brass fittings. *J. Am. Water Works Assoc.*, 80, 45–56.
17. LISK, D.J. (1972). Trace metals in soils, plants and animals. *Adv. Agron.*, 24, 267–320.
18. SCHROEDER, H.A. (1971). Losses of vitamins and trace minerals resulting from processing and preserving of foods. *Am. J. Clin. Nutr.*, 24, 562–73.
19. DABEKA, R.W. and MCKENZIE, A.D. (1988). Lead and cadmium levels in commercial infant foods and dietary intake by infants 0–1 year old. *Food Add. Contam.*, 5, 333–42.
20. JORHEM, L. and SLORACH, S. (1987). Lead, chromium, tin, iron and cadmium in foods in welded cans. *Food Add. Contam.*, 4, 309–16.
21. VARO, P. and KOIVISTOINEN, P. (1980). Mineral element composition of Finnish foods. *Acta Agric. Scand.*, *Suppl.* 22, 165–71.
22. KOLJONEN, T. (1975). The behaviour of selenium in Finnish soils. *Ann. Agr. Fenn.*, 14, 240–7.
23. VARO, P., ALFTHAN, G., EKHOLM, P., ARO, A. and KOIVISTOINEN, P. (1988). Selenium intake and serum selenium in Finland. Effects of soil fertilization with selenium. *Am. J. Clin. Nutr.*, 48, 324–9.
24. BUNKER, V.W., LAWSON, M., STANSFIELD, M.F. and CLAYTON, B. (1988). Selenium balance studies in apparently healthy and housebound elderly people. *Br. J. Nutr.*, 59, 171–80.
25. PENNINGTON, J.A.T. (1988). Aluminium content of foods and diets. *Food Add. Contam.*, 5, 161–232.
26. SHERLOCK, J.C. and SMART, G.A. (1986). Thallium in foods and diet. *Food Add. Contam.*, 3, 363–70.
27. LINDSAY, D.G. (1986). Estimation of dietary intake of chemicals in food. *Food Add. Contam.*, 3, 71–88.

28. WORLD HEALTH ORGANISATION (1973). *Energy and Protein Requirements: Report of a Joint FAO/WHO Expert Committee, WHO Tech. Rep. Ser. No.* 522 (Geneva).
29. WORLD HEALTH ORGANISATION (1985). *Guidelines for the Study of Dietary Intakes of Chemical Contaminants. WHO Offset Pub. No.* 87 (Geneva).
30. SMART,G.A. and SHERLOCK, J.C. (1985). Chromium in foods and the diet. *Food Add. Contam.*, 2, 139–47.
31. ENGLISH, R.(1981). The Market Basket (Noxious Substances) surveys: zinc content of Australian Diets. *J. Food Nutr.*, 38, 63–5.
32. LOUEKARI, K., JOLKONNEN, L., and VARO, P. (1987). Cadmium intake by Finnish Men. *Food Add. Contam.*, 5, 111–17.
33. VERSIECK, J. and CORNELIS, R. (1989) *Trace Elements in Human Plasma or Serum*, 2 (CRC Press, Boca Raton, Florida).
34. SCHWARTZ, K. (1977). *Clinical Chemistry and Chemical Toxicology of Metals*, ed. BROWN, S.S., 3 (Elsevier, Amsterdam).
35. RUSSEL, L.H. (1978). Heavy Metals in Foods of Animal Origin, in *Toxicity of Heavy Metals in the Environment*, ed. OEHME, F.W., 3–23 (M. Dekker, New York).
36. PHIPPS, D.A. (1976). *Metals and Metabolism* (Clarendon Press, Oxford).
37. BOROCZ-SZABO, M. (1980). Effects of metals on sensory qualities of food. *Acta Alimentaria*, 9, 341–56.

CHAPTER 2

HOW METALS GET INTO FOOD

METALS IN SOIL

Soil is the principal source of the metals we find in plants. All the nutrients a plant needs for growth, with the exception of carbon and oxygen, are drawn from the soil. Soil is an exceedingly heterogeneous mixture. It consists of a solid phase, composed of mineral matter, originally part of the rocks of the lithosphere, and of organic matter made up of plant and animal debris in various stages of decomposition. It also has a liquid phase, the soil solution, containing dissolved salts, organic compounds and gases. There is, in addition, a gaseous phase, the soil atmosphere, which occupies those spaces between the solid particles which are not filled with water. Besides these three phases there is a microbial population living in the film of water which surrounds the soil particles and the roots of plants. Bacteria, fungi and algae play an important role in making the soil suitable for plant growth. In this they are assisted by the plants themselves, which secrete organic compounds such as sugars and amino acids which serve as nutrients for the microorganisms.

To grow crops satisfactorily, soil must contain, as a minimum, an adequate supply of all the nutrients essential for plant growth. All soils will contain, in addition, many other elements, generally in very small amounts, in both organic and inorganic forms; some of these elements will be beneficial, others potentially harmful to plants. There is great variation between the types and quantities of trace elements in different soils. This is true even where the soil is ideal for agricultural use and has not been polluted by industrial debris or other contaminating material.

Whatever the actual composition of the soil may be, the plant has to be able to absorb nutrients and carry them to where they are needed in its tissues. The soil liquid phase is the most important immediate source of

24

these nutrients. It is a very dilute solution and would be quickly depleted of its contents were these not constantly replenished by the release of elements from the solid phase. Metals are detatched from the soil reserves partly by solubilisation of minerals and organic matter and partly by ion exchange.

The needs of an actively growing plant could not normally be met by the nutrients supplied by the soil solution found in the immediate area below the stem. It is necessary for the plant to seek out adequate supplies of nutrients, even at some distance from its growing position. This it does through the development of roots as absorbing organs. Roots have evolved in such a way that they can penetrate far out into the soil, pushing through, at times, considerable opposition to bring the greatest possible absorbing surface into immediate contact with the soil matrix. Dittmer's often quoted study[1] illustrates how extraordinary can be the extent of such contact between plant roots and soil, even in a modestly sized annual plant. He found that a single plant of rye, *Secale cereale*, a domesticated member of the grass family, grown for four months in a box of soil just under 56 cm deep and 30.5 cm^2 in surface area, developed a root system whose soil contact area was 639 m^2 and combined length 623 km. Another study found that the roots of a single plant of corn (*Zea mais*) penetrated to a depth of more than 6 m and extended horizontally as much as 10 m within 14 weeks of planting[2]. Undoubtedly, then, when the question of uptake of minerals from soil by plants intended for use as human food is being considered, the nature of both the underlying soil and the surrounding area, not just the top few centimetres of earth, should be taken into consideration.

ACCUMULATION OF METALS BY PLANTS

There is another complicating factor which is of importance when metal contamination of plant foodstuffs is being considered. While all plants will, if placed in a nutritionally balanced soil, take up nutrients to the extent they need for growth, some others possess a special ability which enables them to take up and accumulate, sometimes to very high levels, certain elements. The legume *Astragalus racemosus* is a notorious example of such a plant. Its ability to take in and accumulate large amounts of selenium in its tissues is well known. Selenium, which we will consider in detail later, is now recognised to be an essential plant and animal trace nutrient. However, it is also highly toxic if consumed in more than minute amounts. Indeed, until recently it was its toxicity, not its nutritional role, that was recognised. Selenium is widely distributed in the earth's crust at concentrations of less than 0.1 mg/kg and most vegetables and pasture plants can absorb, without any sign of toxicity, up to about 5 mg/kg of the element. However, certain alkaline soils in Dakota and elsewhere in the USA, have high levels of selenium. This prevents the growth of most grasses and other fodder plants, but not of *Astragalus*, which grows well and accumulates remarkable levels of the element. Concentrations of 15 g of selenium per kg of plant tissue have been reported[3]. If domestic animals, especially horses and cattle, feed on these naturally enriched plants they develop 'alkali disease' or 'blind staggers', a severe and eventually fatal illness. A similar illness in farm animals, due to consumption of another selenium accumulator, has been reported from Australia[4].

GEOBOTANICAL INDICATORS

Plants such as *Astragalus* which can tolerate high levels of metal in the soil and, in many cases accumulate it to toxic levels, are known as 'indicator plants'. Many modern day prospectors consider geobotany, the study of such plants and their association with mineral deposits, a valuable addition to their expertise. Since these plants grow on mineralised soils where other plants may have difficulty in surviving, they can be used as pointers to the presence of local mineral deposits. Many such plants are known. A species of basil, *Becium homblei*, known to prospectors in parts of Africa as the 'copper flower', has often been used as an aid in their searches for minerals[5]. Other indicator plants for a variety of minerals, including uranium, cobalt, gold and silver, have been reported[6].

Indicator plants are of little more than marginal interest, however, with regard to the problem of metal contamination of food. Of greater interest are those strains of normal pasture plants and food crops which develop an ability to grow on metal-enriched soil[7]. These plants are of special significance where regenerated mine dumps or former industrial sites are used for agricultural purposes. They must also be taken into account, as will be seen in a later section, when sewage sludge and recycled refuse is applied as dressing to soil.

METAL ACCUMULATION IN PASTURE PLANTS AND CROPS

Seeds of most plants, if they fall on soil contaminated with high concentrations of copper, lead, zinc or other metals, will fail to germinate or, if they do, will die soon after sprouting. But because of the great genetic variability that exists among populations of plants, there will be occasional individuals with the ability to survive and even prosper in such toxic soil conditions. The resulting plants may not be quite as vigorous or as well formed as if they were growing on normal uncontaminated soil, but they will survive and develop, particularly as they do not have to compete for space and nutrients with other more vigorously growing plants. Extensive studies have been carried out in the UK on strains of the grasses *Agrostis* and *Festuca* which can grow on mine tailings rich in toxic metals[8]. When grazed by animals these grasses can cause poisoning.

In none of the examples given so far of metal accumulator plants is there any real danger of further transfer of the metals to humans[9]. Those animals which die after eating fodder contaminated by lead or other toxic metal are unlikely to be used as food for humans.

However, if the plants are used directly as food, then the problem may become serious. The tragedy that struck the small community of Toyama City in Japan in the mid 1940s was due to the uptake of cadmium by rice growing in paddy fields which were irrigated by water contaminated by effluent from a local zinc–cadmium–lead mine. The cadmium-contaminated rice caused the deaths of more than 50 people and serious illness in many others.

Cadmium contamination still occurs in Japan as well as in other countries. As in Toyama, the source of the contamination is often waste from mineral workings. This was the case in the UK where high levels of cadmium discovered in soil and in locally grown vegetables in the village of Shipham, Somerset, were found to be a legacy from old mine operations[10]. Often such exposure to mine waste leads to multiple contamination of food since ores may contain more than one metal species. In one study carried out in the mineral-rich Copper Belt of Zambia of the effects of wind-borne contamination from a lead–zinc refinery, locally grown corn (*Zea mais*) was found to contain 0.9 mg/kg cadmium and 28 mg/kg lead. Leaf vegetables had on their outer surfaces over 300 mg/kg lead, 20 mg/kg copper, almost 900 mg/kg zinc and 5 mg/kg of cadmium[11]. The health implications of such contamination of food plants are considerable.

It must be stressed that high levels of metals in foods are by no means always due to man's polluting activities. Even apart from the well-known case of accumulator plants, there are many other examples of foods, both plant and animal in origin, that concentrate a variety of metals to potentially toxic levels. The higher than average levels of mercury in the blood of Greenland eskimos has been traced to regular consumption of mercury-contaminated seal meat. The mercury probably originates in the fish and other marine organisms which live in naturally-contaminated arctic waters and are consumed by the seals[12].

Volcanic regions of the world can pose a particular problem with regard to uptake of toxic metals by plant foodstuffs. Volcanic soils are often fertile and, depending on climatic conditions, are often intensively cultivated. This is the case in several parts of the island of Java in Indonesia where a recent study[13] has found high levels of mercury and other toxic metals in soil and water, and considerable accumulation of mercury in locally grown vegetables. It is estimated that consumption of as little as 100 g of potato each day would give an intake almost twice the WHO tolerable level of intake.

A significant source of aluminium in the diet is tea. The shrub *Camellia sinensis*, like a number of other plants, has the ability to absorb the metal from the soil solution and concentrate it in its leaves[14]. These are the leaves which are dried to make the tea which is used to prepare what is probably the most widely used beverage in the world. Whether the enormous quantity of this aluminium-containing beverage which is drunk by about half the world's population constitutes a danger to health seems unlikely. The plant also has the ability to accumulate the essential element manganese in the same way. It is believed that tea drinking makes a significant contribution to the manganese intake of many elderly persons in the UK[15].

Several other examples could be given to illustrate the fact that not all toxic metal uptake and accumulation in foods is due to accidental contamination or to pollution of the environment by human activities. We will look at some of these incidents when considering the individual metals in later chapters. The examples given here should be sufficient to illustrate a basic fact which is that metal accumulation in foods is a natural occurrence, with benefits as well as possible disadvantages to those who consume the foods. This fact will be taken for granted in this book, but special attention will be given to those instances in which metal accumulation is sufficient to constitute a health hazard.

METALS IN SEWAGE SLUDGE

Of considerable significance, from the point of food contamination, is the practice of applying sewage sludge to agricultural land as a top dressing. The modern world produces an enormous amount of waste as a result of both domestic and industrial activities. When needs were simpler and the world was not so crowded, there was little difficulty in disposing of refuse. But now, not only is the volume of waste constantly growing, but with increasing use of a wide range of harmful substances including exotic and often toxic metals, the waste we produce requires costly and often highly sophisticated handling.

An idea of the size and composition of domestic waste generated by a modern community is given by data from the relatively small Australian city of Brisbane, Queensland[16]. The city has a population of some 750000. Total domestic refuse delivered to all the municipal dumps over one year was estimated to weigh 600000 tonnes, or about 1.4 kg per person each day. The total garbage dumped, apart from automobile bodies and other large metal objects which were salvaged for recycling, was estimated to contain 29000 tonnes of steel and 5500 tonnes of aluminium. No record was made of quantities of other metals which were also found in the refuse. No doubt other larger and more industrialised communities produce even greater amounts and varieties of domestic waste than does Brisbane.

Much of this urban waste is deposited as landfill in holes in the ground, such as disused gravel pits, mine excavations and even natural hollows. Some is dumped at sea. An increasing amount is incinerated and the resulting ash used as landfill or as building aggregate. A considerable amount of our waste is converted into sewage sludge and municipal compost in treatment plants in some countries. In England and Wales, for instance, over a million tonnes (dry weight) of such sludge is produced each year. Quantities produced in other technologically developed countries are equally impressive.

The idea of using sewage sludge as a dressing on agricultural land and thus converting 'the refuse from your garbage can into plant food' has considerable economic and ecological attractions, as well as political value in a world that is becoming increasingly eco-conscious. There is much to commend sludge to the farmer as a top dressing for his arable land. It is rich in organic matter, on average about 40 per cent of its dry weight. In addition it contains about 2.4 per cent nitrogen and 1.3 per cent phosphates. Modern farming methods can lead to unstable structure in some types of soil which can be rectified by the addition of suitable organic matter. Sludge is ideal for this purpose. It is also a cheap source of both nitrogen and phosphorus, two essential and normally expensive plant nutrients.

Unfortunately, in addition to its desirable qualities, sewage sludge has a negative side. It can contain relatively high levels of a variety of different metals. Sometimes these are in concentrations sufficient to make soil to which sludge is applied toxic to plants. Incidents of crop failure due to this cause have been reported[17]. The metals in sludge usually are industrial in origin, but domestic waste can also make a substantial contribution. Mercury, zinc, lead, cadmium and other metals have all been reported to occur, sometimes at high levels, in household dust[18] as well as in domestic garbage[19]. Boron, for instance, seems to come from washing powders, some of which contain as much as 1 per cent of the element. Zinc has been attributed to the use of galvanised iron in structures and equipment of the home.

Table 7 indicates the range of concentrations of several metals which can be expected to occur in typical sewage samples. These are considerably higher than levels found in normal agricultural land, with more than 300 times as much zinc and 100 times as much boron and copper as would occur in arable rural soils[20].

When such material is added in even moderate amounts to agricultural land, a considerable increase in levels of toxic and other metals can be brought about. The added metals, apart from boron, are not easily leached out again. As a result, after even one treatment, without further addition of sludge, the metals can continue to have an effect on crops for many years. Studies by Purves and his colleagues in Edinburgh have shown, however, that while such enrichment does result in an increased uptake of metals by certain types of plants, not all the metals are taken up to the same extent. In addition, there can be considerable differences between levels of uptake by different species of crops as well as by different tissues of an individual plant.

Differential accumulation of metals in different parts of a plant is significant when potential risk to consumers is being considered. Purves and his colleagues found, for instance, that while boron was the element most readily accumulated by most plants, in the case of beans at least, most of it was concentrated in leaves and stems, not in the seeds. This reduces considerably the potential health hazard of the uptake. Copper was also found to be concentrated in leaves, but to such a limited extent that even in leafy vegetables such as lettuce, there could be little danger to

TABLE 7
Normal Range of Metals in Dry Matter of Sewage Sludge[21, 22]

Metal	Content (mg/kg)
Boron	15–1 000
Cadmium	60–1 500
Chromium	40–8 800
Cobalt	2–260
Copper	200–8 000
Iron	6 000–62 000
Lead	120–3 000
Manganese	150–2 500
Mercury	3–77
Molybdenum	2–30
Nickel	20–5 300
Scandium	2–15
Silver	5–150
Titanium	1 000–4 500
Vanadium	20–400
Zinc	700–49 000

the health of consumers. Zinc accumulation in leaf tissue of a variety of vegetables was also well below toxic levels.

The finding of less than toxic levels of some metals in crops grown on sludge-enriched soil was confirmed by Pike and colleagues[21]. They studied arable land which had served for several decades as a sewage farm and was later used for general agricultural purposes. They found that while levels of copper, zinc and nickel in soil of the farm were generally higher than in normal agricultural land, uptake by lettuce and radish of copper and nickel was no greater than from normal soils. However, significant amounts of zinc were accumulated by the crops, but this accumulation was unlikely to be a major health problem for consumers because of the low toxicity of the element. The chemical form of toxic elements in sludge-amended agricultural soil is also of interest, since uptake and toxicity of some elements depends on their form. Capon[22], in a study of vegetables grown in residential garden plots treated with sludge from a suburban sewerage treatment plant, found mercury and selenium levels four and two times those in crops grown on untreated soils. Fourteen per cent of the mercury in the treated plants was methylmercury compared with 4.4 per cent in the control crops.

Lead, cadmium and mercury in waste-derived sludge have been found to give rise to much more serious problems than other metals such as copper and zinc. Purves and others have drawn attention to the special problems presented by these elements in sludge-treated soil. These problems have been recognised by the Swedish National Board of Health and Welfare which prohibits the use on agricultural land of sewage sludge containing more than 15 mg/kg of cadmium[23].

The Toyama tragedy in Japan, where cadmium from an industrial source was released into irrigation water and resulted in contamination of food crops, was one of the most dramatic and widely publicised accidents of this nature. Many other incidents have been reported since then in several parts of the world. It is fortunate that the earlier incidents served to alert the community and triggered much activity of environmental protection agencies and other watchdog bodies in many countries. However, similar accidents will, undoubtedly, continue to occur. Some may be even more serious, or less easy to detect and control, than Minamata. There is considerable potential in modern industry for environmental and food chain contamination. A wide range of potentially toxic metals is employed by industry and these can, and do, find their way into sewage sludge and municipal compost. Many have been widely used for centuries, others have only recently found industrial applications and their potential as health hazards is not yet widely recognised. We will consider some of these 'newer' hazards later. Health authorities and the food industry must continue to exercise vigilance in order to protect the community from the unexpected risks to health which can result from the introduction of new materials and the development of new techniques which generate waste capable of entering the food chain. This book is designed to help those with responsibility for protecting the food supply and the right of consumers to pure food, to anticipate such situations and to be prepared to handle them.

METAL PICK-UP BY FOODS FROM AGRICULTURAL CHEMICALS AND FERTILISERS

A report of the National Environmental Protection Board of Sweden[23] commented on the fact that, while concern was being expressed by some authorities on the danger of using sewage sludge in agriculture, the far more serious problem resulting from the presence of toxic metals in commercial fertilisers and agricultural chemicals was being overlooked. For example, while the Swedish National Board of Health and Welfare restricted the use of cadmium-containing sludge to a level which resulted in the application of not more than 15 g of the metal per hectare of treated farm land, some widely used commercial fertilisers could introduce much higher levels of cadmium into the soil. Thus one type, NPK 16–7–13, was found to contain 18 mg/kg of cadmium; several others had more than 20 mg/kg and one superphosphate preparation had 30 mg/kg. The report showed that while the overall use of sewage sludge for soil improvement in Sweden in 1973 was about 84000 tonnes (equivalent to 1260 kg of cadmium if levels were at the permitted maximum levels) total use of commercial phosphate fertilisers was about 732 000 tonnes, equivalent to approximately 10 000 tonnes of cadmium. This is more than 10 times the total contribution made by sewage sludge. The report also noted that lead levels in fertilisers were high, with approximately 50 tonnes of the metal spread on Swedish agricultural land each year.

Similar findings of heavy metal contamination of soils and crops resulting from the use of fertilisers have been reported from Australia[24]. Uptake of cadmium especially by wheat from soil which has been contaminated by fertilisers can be particularly high, but other food crops can also be seriously affected.

A factor which illustrates the complexity of the problem of metal contamination of food crops is reported by Kjellström and his colleagues working in the Department of Environmental Hygiene, Sweden[25]. They point out that it has been shown that an increase in soil acidity results in increased availability of cadmium for uptake by plant roots. A major cause of lower soil pH is 'acid rain' resulting from atmospheric pollution largely resulting from the combustion of fossil fuels. This is a world-wide phenomenon, but is of particular concern in Scandinavia where prevailing winds bring combustion products from Britain and elsewhere in western Europe. Not just dying conifer forests but, as the authors of the report point out, a reduction of soil pH results from this pollution. A consequence of this is increased availability for uptake by crops of cadmium, even when soil levels of the element remain constant.

Increases of levels of metals in soil and their uptake by food plants can have beneficial as well as undesirable effects. In fact, deliberate manipulation of soil conditions on a national scale has been undertaken in Finland and is being considered in other countries as a means of overcoming what is seen as a public health problem. Studies in the 1970s showed that dietary intake of selenium in Finland was considerably less than in some other countries, particularly the US. Locally produced foodstuffs had low levels of the element and daily intake of an adult was estimated to be approximately 50 µg/day, about half the US intake. In some individual cases, levels of intake were comparable to those found in regions of China where Keshan disease, a cardiomyopathy related to selenium-deficiency, occurred. Concern was expressed by Finnish health authorities that chronic ingestion of such low levels of selenium could cause considerable health problems, including heart disease and even cancer, and consideration was given to steps which would improve national intake. As will be described in some detail later, this was eventually achieved by enriching fertilisers used in the country with selenium. The result has been spectacular, with levels in most locally produced foods rising considerably and average intake increasing from what was among the lowest in the world to approximately 100 µg/day[26]. Whether this has, in fact, the expected health effects, remains to be seen.

AGROCHEMICALS

Several metals are used, in inorganic as well as organic forms, in fungicides and other agricultural chemicals. One of the earliest to be used was copper, as copper sulphate mixed with lime and water in Bordeaux mixture. It was sprayed on grapes, it is said, originally to deter stealing before the fruit could be picked for wine-making, but was later found to prevent mildew. Potato growers know it as an important preventative of blight and other fungal diseases of their crops. It is still used widely on a number of crops subject to fungus attack. If improperly applied, for example in excessive quantity or too near harvest time, such copper-containing preparations can cause food contamination. Since potatoes are more often than not peeled before eating and the spray is applied to the haulms above ground and not on the tubers themselves, Bordeaux mixture is unlikely to pose a serious danger for the consumer. High levels of copper sometimes found in wines, however, can occasionally be traced back to late spraying of grapes.

The use of organic mercurial compounds has been the cause of far more serious and better documented cases of food poisoning. These compounds were once widely used as antifungals for seed dressing. The misuse of treated seed wheat for flour and bread-making caused an epidemic of mercury poisoning in Iraq in the late 1960s[27]. Similar misuse in Pakistan and Guatemala a decade later again caused poisonings of epidemic proportions. Mercurial compounds are no longer common in agricultural practice and have been largely replaced by less persistent fungicides. However, mercury-containing agrochemicals are still available, as are those containing other toxic metals and they still can on occasion make a contribution to dietary intake.

Arsenic is used as an insecticidal spray in orchards to control such pests as ermine moth on apples and the sawfly on pears. The most common form of the element is in combination with another toxic element, lead, as in lead arsenate. Normally such compounds are applied well before harvest time but, on occasion, the time between spraying and picking is too short and the fruit carries arsenic and lead to the consumer. This has been a problem with some imported fruit and has resulted in the condemnation by UK customs officers of some batches of Australian apples. It probably also accounts for high levels of both elements reported to occur sometimes in cider and perry.

A recent survey of arsenic in food published by the UK Ministry of Agriculture, Fisheries and Food[28] noted that while use of arsenic pesticides in horticulture has declined appreciably, the use of lead arsenate was still permitted. This use, the report indicated, may be responsible for the occasional finding of significant levels of arsenic in canned apple baby food and other apple and pear products. It was also noted that fruit and vegetables are the third most important source of arsenic in the diet, possibly again reflecting the use of arsenic compounds in orchards and market gardens. The two highest sources of the element in the diet were, first, seafood and, second, animal fats. While most of the arsenic in fruit and vegetables probably originated in agricultural chemicals and soil, its presence in seafoods may largely be due to industrial pollution[29].

Organic arsenic is used as a contact herbicide and, like any similar compound, is capable of entering the food chain and causing contamination if misused. Another compound of arsenic, arsenilic acid, is sometimes added to animal feed as a growth promoter. The amounts used are very low and it is unlikely that such growth promoters will increase significantly the level of the element in human foods. However, again, abuse or accidents could lead to isolated incidents of contamination.

CONTAMINATION OF FOOD BY METAL-CONTAINING WATER

Man is a profligate user and misuser of water. Waste of many kinds, from raw sewage to industrial runoff has been allowed to pour into rivers, lakes and seas until sometimes the waters become corrupt and sources of disease. All the great cities of the world experienced, at some time or another in their histories, the foul smells and the diseases bred in polluted waters. But eventually realisation grew among those with power to act, that pollution and the resulting sicknesses were preventable. In the middle of the nineteenth century great sewage and waterworks schemes were undertaken in many countries. With the implementation of stringent legislation, many once foul rivers were returned to a state approaching that which they enjoyed before the great population and industrial explosions took place. Salmon and trout began to appear once more in the great rivers of Europe. Sometimes, unfortunately, their stay was not prolonged. Nevertheless, the achievement was great.

But just, it might seem, to remind us that congratulations might be premature, the Minamata tragedy occurred in Japan. Mercury used as a catalyst in a factory manufacturing plastic was discharged into what had been a small, clean bay of the Sea of Japan. Pollution occurred, but its effects were, for a long time, unnoticed. Insidiously, the metal was spread out on the bed of the sea where it was picked up and accumulated by the minute organisms that lived there. The inorganic mercury underwent a transformation in the bacteria and other minute creatures into an organic form. Larger organisms fed on the mercury-containing smaller creatures, and they in their turn were consumed by still larger organisms, and at each step the organic mercury compounds were handed on to new hosts. Eventually, the metal, now in fish, became the food of the fisherman of Minamata and their families. The result was the horrifying epidemic of organic mercury poisoning which brought the name of Minamata to the attention of the world. We were reminded that pollution of once clean waters still occurred, with disasterous effects on human health.

The subsequent discovery of organic mercury in fish in American and Swedish lakes showed that mercury contamination of water was not confined to Japan alone. It was found that large amounts of mercury were being released into freshwater lakes and rivers by a variety of industries in North America and levels of the element were found to be increasing in fish caught in the Great Lakes and other waters of the several Canadian states[30]. Similar discoveries were made in Sweden. The culprit for much of this contamination was found to be the wood pulp and paper-making industries. It became clear that in these countries mercury-polluted waters were the cause of significant contamination of the human food chain.

A comment on the relation between quantity of a contaminated food consumed and the risk involved is worth making at this point, since an important toxicological consideration is well illustrated by such incidents. As was noted by the authors of a report on one of the Canadian incidents[31], in which mercury was released into a lake from a chloralkali plant, highest blood and hair mercury levels were found in fishing guides in tourist camps and in members of their families. While these people, who consumed fish regularly and in relatively large quantity, were at risk of mercury poisoning, visitors and others, whose intake of the contaminated food was low, were unlikely to suffer adverse effects from the metal. Total quantity consumed, as well as the length of total exposure to a contaminated foodstuff, must always be taken into account when risks of this nature are being considered.

Polluted water has also been the cause of cadmium poisoning through contamination of food. The Toyama City incident in Japan has already been mentioned. In Sweden, too, similar accidents have occurred. For instance, fish downstream from a nickel–cadmium battery plant have been found to contain toxic levels of both metals. In the US, as was shown by a Federal Geological Survey in 1970, 4 per cent of all samples collected from rivers and reservoirs all over the country had cadmium concentrations exceeding the maximum allowable levels for drinking water.

Large-scale pollution of water by other metals besides cadmium and mercury has been shown to occur in many industrialised countries. Typical of such incidents involving works effluents are the levels and variety of metals released into the River Tees in the north-west of England by a steel-making plant. The effluent was shown to contain between 10 and 15 mg/litre of dissolved chromium, copper, zinc, aluminium and manganese as well as 30 mg/litre of iron and lesser amounts of arsenic, uranium, titanium, vanadium, cobalt, tin, zirconium, antimony and tungsten, a varied and potentially lethal cocktail[32]. It was estimated that the South Teeside steelworks released each day 7500 kg of iron, 2300 kg of manganese, 850 kg of zinc, 310 kg of lead and 25 kg of copper. This was mainly as particulate matter, not in soluble form, and most settled out on the bottom of a holding lagoon before the effluent was released into surface waters. Changes in industrial practice and the introduction of stringent environmental protection laws has changed the picture remarkably in the past decade and such pollution incidents are rarely met with in the UK, except as the result of accidents. Crucial steps were taken in the early 1970s to limit pollution of waters with the introduction of the Deposit of Poisonous Waste Act (1972) and the Control of Pollution Act (1974). These and subsequent Acts and numerous local authority regulations have had considerable success in reducing industrial pollution of water systems.

These anti-pollution laws and regulations have been especially effective in controlling an industry which has traditionally been a major generator of toxic effluent and has contributed a great deal of metal contamination to water systems. Metal-finishing produces great amounts of toxic and corrosive solutions and metal-rich sludges. Those produced by plating shops especially can be very complex and variable in composition. Typical of the analysis of effluent from a metal finishing plant is the following: Nickel 2.5 per cent, zinc 1.5 per cent, chromium and copper 1 per cent, cadmium 0.25 per cent, lead 0.2 per cent[33]. Discharge of such effluents into sewage systems or into rivers and other water systems is, since the passing of the Acts, a costly business. Economic, no less than social pressures, have led the industry to develop new methods of handling the waste it produces. These include detoxification of discharge and recovery of metals and other valuable waste products. With public opinion increasingly more vocal and politically powerful, and environmental protection agencies so ready to pounce when what have been called 'environmental pollution highlights associated with the disposal of industrial waste' occur, the old 'out of sight, out of mind' approach to disposal of toxic by-products is no longer acceptable.

Contamination of food by metals in industrial effluent and industry-polluted water is, fortunately, not common. The Japanese incidents of mercury and cadmium poisonings in the 1950s and 1960s, as well as evidence provided by the presence of polluted crustaceans in the coastal waters of many countries, show that responsible authorities must always be alert to the dangers that necessarily result from our present industrialisation. It must be remembered that it is a controllable problem where good industrial housekeeping and the operation of well designed municipal and other water-treatment plants can remove the danger of excessive levels of metals in water used for food production and drinking.

METAL CONTAMINATION OF FOOD DURING PROCESSING

Contamination at the Factory Door
It would be misleading if the discussion in the foregoing section of possible sources of metal contamination of primary foodstuffs was to give the impression that it is normal for a high proportion of foodstuffs arriving at processing plants to contain more than minute amounts of toxic or potentially toxic metals. In fact, it is rare for this to occur. Current legislation and the monitoring practices of public health inspectorates in most of the developed countries see to it that levels of contaminants in primary food products processed for sale are normally at, or below, levels which are known, in the US, as GRAS — 'generally recognised as safe'. However, most vegetables and many other primary foodstuffs will require some form of cleaning before processing. At the factory gate they may contain particles of soil, dust and other 'dirt' picked up during growth, harvesting and transport. Under certain conditions, to which reference has already been made, dust and soil particles may be rich in metals. If these are not removed they could cause contamination of the processed food.

Contamination on a gross scale has been known to occur from the presence of rogue metal — such as fragments of harvesting or processing equipment which have by accident found their way into the food. Foods are routinely screened for such unwanted materials by one or a combination of standard cleaning operations carried out as a first step at food processing plants. Screens are employed for sifting out larger particles and various types of metal detectors are used to pick up fragments that escape the sifting. In-line controls can be set up which are capable of detecting the ferrous and non-ferrous metals most frequently encountered as contaminants of this type. Such equipment and procedures are covered in texts on food engineering where details will be found[34].

No matter how free from metallic contamination food is before it reaches the processing plant and kitchen, it is still possible that the end-product which is served or sold to consumers could contain undesirable amounts of certain metals if care and certain precautions are not adopted.

Contamination from Plant and Equipment
Food processing equipment and containers have long been recognised as sources of chemical as well as microbiological contamination of food. It is said that the Romans suffered from chronic poisoning by lead leached

from the glazed pottery vessels they used to store wine[35]. In more modern times use of similar vessels to hold pickled olives has resulted in lead poisoning in Yugoslavia[36]. Lead has always been a problem in food processing because the metal or its alloys readily lends itself to the fabrication and repair of cooking and storage utensils. The use of strips of lead sheet to repair cracks in wine casks, a not uncommon practice in certain French vineyards, has been shown to cause lead contamination of wine. As long ago as the eighteenth century cider drinkers in England sometimes suffered from 'Devonshire dry gripes', the colic and stomach cramps of lead poisoning, caused by the metal picked up from the lead-lined troughs used in making the beverage[37].

Amateur repair work on cooking and storage vessels, in which lead solders are used, are relatively common causes of food contamination. However, the consequences are usually of limited significance and more often than not occur in small-scale and home production rather than in commercial processing plants. However, use of inappropriate materials for repair of equipment does occur at times even in sophisticated manufacturing plants. The replacement of a corroded, stainless steel pipe in a freezer unit by a copper tube salvaged from a discarded piece of equipment in a large Australian ice cream plant resulted in copper-contaminated products and several poisoning reports[38].

Metals for Food Processing

In modern food-processing plants which have been designed with full regard for the needs of hygiene and cleanliness in mind, taking the types of materials being processed into account, and with awareness of their compatibility with structural materials to be used, there is little danger of metal pick-up by the products. High-quality stainless steel, plastics and other structural materials approved for contact with foods are used. The actual grades of materials used will depend on the particular products to be handled. Dairy products, for instance, are particularly sensitive to some forms of contamination and require special consideration. A full treatment of metals and alloys suitable for fabrication in the dairy industry is given by Harper and Hall[39] and other food-processing industries are well covered in the text by Brennan and his colleagues referred to above[34]. These should be consulted for further details on engineering aspects of food-processing plant design. Here we will look at the composition of some of the steels and other alloys and plating metals used in the food industry, since these are the source of some of the contamination in food caused by the individual metals we will consider in later chapters.

Whitman[40] has discussed interactions between structural materials used in food-manufacturing plants, and foodstuffs, cleaning agents and other substances, and the likelihood of their causing food poisoning or quality defects in products. He pointed out that arsenic, lead, mercury, cadmium and zinc are all potential causes of food poisoning. In addition copper, nickel, iron and chromium, even in very small amounts, can act as catalysts in oxidation reactions and result in rancidity of fats and fat-containing foods. All of these metals may be present in stainless steel, so it is important that the quality of the steel used in food processing be high and good operating procedures be adopted to reduce to a minimum the possibility that these contaminants will migrate from equipment to the food during processing. An important factor in relation to metal pick-up by foods is the length of time they are in contact with equipment. A study[41] carried out at the Institute of Hygiene, Rome, found that while manganese, chromium, iron and nickel did migrate from stainless steel into food, the level of migration was relative to contact time. With high quality stainless steel, even after 30 days of contact, the amount of trace metals migrating into food was very small and well below toxic limits. However, from the nutritional point of view, it is interesting that migration of chromium from stainless steel cooking and processing equipment could be an important source of this essential trace element in the diet.

Unsuitable Metals for Food Processing

The problem of metal contamination during food processing occurs frequently as a result of misuse of equipment or the overlooking of the consequences of using unsuitable metals in apparently insignificant ways. One example of this has been given above in which a steel pipe was replaced by a copper tube in an ice cream plant. Cadmium, which is often used in engineering to plate small pieces of equipment, is readily soluble in weak acid solutions. Its use in food processing plants should be confined to parts which do not come directly into contact with food. Unfortunately, cadmium-plated vessels have been used improperly on occasion and poisonings have resulted. Nickel also is a useful plating metal and, as with cadmium, can be dissolved if it comes into contact with food. Nickel is not considered to be a toxic metal, but it is a very effective catalyst. It can cause oxidative spoilage and rancidity in fat-containing foods with which it comes into contact. Scraped-surface heat exchangers are often nickel-coated. If they are moved in a food plant to a different duty from that for which they were designed originally, corrosion can occur and contamination of food products result.

Copper, like nickel, catalyses rancidity of fats and is a frequent source of trouble in the food industry. The metal itself, or brass, or any other copper-containing alloy, should not normally be allowed to come into contact with fat. Tinned wire baskets made of brass have been implicated in copper contamination of oils used in deep-fat frying. Even brief use of a thermometer encased in a copper tube to test the temperature of heated cooking oil was found to increase the copper content of the oil fourfold, with resulting rancidity. Brass and bronze plumbing fittings have been shown to result in copper, lead and zinc contamination of water as well as of wine[42].

However, not all use of copper or brass equipment in the food industry results in contamination. Copper heating coils and cooking pans are, in fact, widely used in food-oil refineries and in confectionery manufacturing plants, without ill effects. This is probably because the metal is not subject to wear under the circumstances of use. In addition, a thin film of cooked material coats the metal surfaces and provides some protection. Indeed, it is not unusual for initial problems of rancidity and contamination to resolve themselves with time. Similar improvement with time has been observed when zinc-plated galvanised iron equipment is used in food processing. With extended use a surface barrier appears to build up to protect the soluble zinc plate.

Contamination of Food in Catering Operations
Tinned copper pots and pans have traditionally been favoured by professional chefs. The combination of metals was found to be ideal for cooking: the copper, with its high conductivity, allowed rapid heating while the tin protected the easily dissolved copper from attack by food acids.

Today such cooking utensils are still to be found in hotel and restaurant kitchens. Though stainless steel, aluminium and a variety of non-stick surfaced metal utensils, as well as ceramics and glass for the microwave oven, have been adopted in the majority of domestic kitchens, tinned copper pans and pots are sometimes even today to be found in home use.

While the good qualities of such tinned copper utensils are still valued, it has also been recognised that their use can cause contamination of food. The tin used to plate the copper is, in fact, not pure tin but a lead–tin mixture. The lead is required for technical reasons as a flux so that the

TABLE 8
Copper and Lead Contents of Food Before and After Cooking (mg/kg)

	Fish		Chicken		Cabbage		Potato	
	Pb	Cu	Pb	Cu	Pb	Cu	Pb	Cu
Uncooked	0·31	0·82	0·14	2·21	0·15	1·36	0·19	3·10
Aluminium	0·36	1·37	0·21	2·52	0·18	1·04	0·16	1·87
Tinned copper (old)	0·42	5·70	0·25	6·36	0·29	2·07	0·22	2·39
Tinned copper (unused)	1·09	2·24	0·94	4·05	0·79	1·93	0·26	1·88

plate can be spread evenly over the surface of the copper. Even though the amount of lead used is very small, it is enough in some circumstances to contribute a significant amount to foods cooked in tinned utensils.

A study carried out on food prepared in traditional saucepans in a commercial restaurant[43] showed the levels of contamination that can occur. Even though the composition of the tin used on the pots and pans conformed to the limit of 0.2 per cent lead allowed by UK Food Regulations, it was found that 8–9 mg/litre of the metal were leached from the surfaces of a series of new tinned saucepans by a 4 per cent acetic acid solution, which is equivalent to normal vinegar. This is the standard test used to determine levels of leaching of toxic metals from catering utensils. In older saucepans, on which the tin layer was becoming worn and would soon need to be sent back for re-tinning, leaching of lead was considerably reduced but, as might be expected, the copper was now dissolved by the solution. Measurement of levels of copper and tin in foods cooked in the old and newly surfaced utensils, as well as in an aluminium saucepan, indicated the amount of leaching that occurred under actual kitchen conditions. It is clear, as Table 8 shows, that both copper and lead were leached to an extent dependent on the state of wear of the utensils. In a similar study[44], a commercial dried tomato soup was rehydrated and heated in a tinned copper saucepan with the following results for lead: previously unused saucepan: 3.53 mg/kg; worn saucepan: 0.71 mg/kg. The tinning on the new saucepan was found to contain 0.28 per cent of lead (above the legal limit) while the plate on the used saucepan had less than 0.1 per cent. As we shall see later in the chapter on lead, it is not uncommon to find high levels of lead, well above legally-permitted limits, in the plating on some cooking utensils sold to the public.

Other metals may also be picked up by food from cooking equipment, sometimes to a toxic level. There is evidence that the African tradition of using cast iron pots, especially over slow fires for long cooking times, has contributed to a high intake of the metal. In some cases intake is so excessive that it may have contributed to iron-overload of the liver observed among certain indigenous peoples in southern Africa[45]. It is probably true that the same situation occurs in other simple societies, as has been reported in Papua-New Guinea by Drover and Maddocks[46]. A connection has been postulated between the use of copper and brass utensils in the preparation of infant foods and Indian childhood cirrhosis (ICC) that occurs in certain parts of that sub-continent[47]. ICC is characterised by deposition of copper in the liver.

As will be noted in a later chapter, there is growing concern that aluminium may be associated with certain degenerative diseases, including Alzheimer's. While there is little evidence to support the view, the release of aluminium into food cooked in utensils made of the metal continues to be postulated by some as a contributing factor in these diseases. Aluminium may, indeed, be picked up, especially by acid foods, from contact with aluminium saucepans, foils and containers[48], but whether this is to toxic levels for normal individuals is doubtful.

While the deleterious results of metal pick-up from cooking and food storage utensils have been stressed here, it is appropriate to recognise that such adventitious inclusions in prepared foods may also have beneficial effects on health[49]. It is probable that much of our necessary chromium intake is provided in this way, as leachate from stainless steel utensils[50]. Stainless steel could also be a source of dietary nickel. The dietary significance of such adventitious iron, zinc, copper and lead in domestically-prepared foods has been discussed at length by Reilly[49].

Just as in industrial food processing, so in domestic and even commercial catering establishments, the misuse of items of equipment can lead to metal contamination of food. Galvanised iron is sheet iron coated with a layer of zinc to prevent corrosion. The metal coat, which often contains small amounts of cadmium mixed with the zinc, is easily dissolved by dilute acids. For this reason galvanised utensils should never be used for holding foods which are acidic in nature. Unfortunately, this has not always been recognised by cooks and manufacturers and, as a consequence, cases of food poisoning have occurred. Galvanised buckets have been used to hold soup and meat stock in kitchens and there is a temptation for the home brewer and wine-maker to use readily available galvanised drums to hold their fermentation mixtures. As a result many home brews have been spoiled through the development of a metallic taste or, even worse, become toxic because of a high content of zinc and other metals[51].

In general, zinc-plated metal should only be used for storage racking, hooks and other items which will not come into immediate contact, or at least have only transitory contact, with wet foodstuffs. An even more restricted use should be made of iron or steel containing certain other toxic components. Cadmium-plated metal has already been mentioned. In addition, beryllium, which is used as a hardener in copper alloys, should be excluded from all possible contact with food. The alloys are highly toxic and, even if initially used in a safe position in a food-processing plant, there is always the danger that subsequently the beryllium-containing item may be used for a different function and food contamination could result.

Ceramic and Enamelled Utensils as Sources of Metal Contamination of Foods

Pottery containers and cooking utensils continue to cause metal contamination of foods in spite of warnings by health authorities of the need for caution when certain types of pottery are used in connection with food. Though the problem is limited in scale and seldom of consequence to food manufacturers, metal contamination of foods due to the use of poor-quality pottery vessels must be included in this study. On occasion, commercially-prepared products are marketed in specially-designed pottery containers as a marketing gimmick. Costly product recalls could be necessary if care is not taken to see that the pottery is properly glazed and product contamination does not occur. A special gift-pack of a well-known Australian rum suffered this fate in recent years.

Pottery used for food containers and cooking vessels is normally glazed to produce a non-porous, water-tight surface. The glass-like glaze of good earthenware and ceramics is produced by coating the surface with a carefully prepared frit and heating it to a high temperature in a kiln. The frit, which contains salts of lead and other metals to act as a flux, is vitrified and forms a glass-like layer on the pottery surface. The glaze on properly made and kilned plates and dishes should normally be unaffected by food and should not release its lead and other metal components during cooking or standing. Unfortunately, some glazed pottery, especially of the craft and home-made kind, which has not been kilned at a sufficiently high temperature or has been made with poorly formulated frits, is capable of releasing toxic amounts of lead, cadmium and other metals into food.

The British Consumer Association found that many of the earthenware casseroles on sale in high-street stores were poorly glazed and capable of contaminating food. A Canadian study of a range of different earthenware items being sold for domestic use found that 50 per cent were unsafe for use with food[51].

Leaching of toxic metals from glazed ceramic surfaces is more of a problem with smaller articles than with large, robust vessels and containers. The harder stoneware from which these items are normally made appears to be less liable to release toxic metals into food than are the casseroles and dishes studied by the British and Canadian investigators. This may be some consolation to food manufacturers who are concerned about the safety of the glazed vats and sinks in their processing plants.

From the domestic user's point of view, the best line of action with regard to pottery utensils is to follow the advice of the UK Ministry of Agriculture, Fisheries and Food[52] and use only glazed ceramics produced by reputable manufacturers. They should leave craft and home-made pottery as ornaments on the shelf.

Because of the continuing problem of leaching of heavy metals from glazed ceramics, authorities in various countries have introduced regulations and guidelines with regard to the quality of such items. The Council of the European Economic Community has issued a Directive laying down limits for lead and cadmium leaching, under specified conditions, from ceramic articles[53]. The standard is similar to that already in force in a number of countries such as Finland[54]. The International Organisation for Standardisation also established similar permissible maximum levels for lead and cadmium release from ceramic ware[55]. The method for testing the ceramic ware accepted internationally uses a 4 per cent acetic acid solution to extract the toxic metals from the surface of the utensils. This is the method used by the Ceramic Dinner Ware Lead Surveillance Program which has been instituted in the US.

Enamelled kitchen ware can also be a source of metal contamination of food. Enamel is rather like the glaze on pottery: a glass-like layer baked on to the surface, serving both as an anticorrosive lining and as a decoration on a metal utensil. As is the case with ceramics, enamelled casseroles, plates, saucepans and other utensils which come from reputable manufacturers and which are guaranteed to conform to British or other standards, will be quite safe to use. But others of unknown origin, whose adherence to safety standards cannot be guaranteed, are suspect. This is especially so if they are bright red or yellow in colour, indicating the possible use of lead and cadmium pigments. Old worn enamelled plates and dishes with cracked surfaces are also suspect, since they have been found to contaminate food not only with the now exposed underlying metal, but also with metals released from the enamel.

Another domestic source of cadmium and lead contamination is the decoration and printing used on food and beverage containers. A Japanese investigation found that some decorated glass drinking glasses were capable of releasing high levels of both metals into beverages. The toxic metals came, not from the glass, but from the pigments used to decorate the glasses[56].

Print and colour applied to plastic containers and wraps have also been shown to be capable of contaminating food. Some polythene bags, for instance, have been found to contain nearly 25 g/kg of lead. Some of this lead comes from the metal salts used as stabilisers in the manufacture of the plastic, but pigments used for colouring, as well as the printing on its surface, are also responsible for the lead, as well as of other metals.

Wrapping paper and cardboard containers also contribute to metal contamination of food. The metal comes both from the paper and card as well as from printing and decoration on the surface. Some paper food wrappers have been found to contain as much as 50 mg/kg of lead. This may be due to the use of recycled printed paper to make card. Letterpress magazine ink may contain nearly 30 g/kg lead in yellow and 4 g/kg in red print. The unfortunate implications for resource recycling and conservation of the environment are obvious.

A study[57] of coloured paper wrappings used for food and confectionery found unacceptably high levels of lead. One sweet wrapper, coloured with a mixture of yellow, blue, red and brown pigments, contained more than 10 g/kg of lead. A candy-bar wrapper was somewhat less toxic with approximately 7 g/kg of the metal. Lesser quantities of lead were found in food packs, with 20 mg/kg in a paper flour bag and 50 mg/kg in a cardboard spaghetti box. While it is unlikely that the flour or pasta packings would find their way into the diet, this is not the case with the candy and sweet papers which are quite likely to be chewed at least by younger children. The possibility of excessive intake of heavy metal from such sources is considerable.

It must be pointed out that the contents are often separated in packs from the printing or decoration by an intervening barrier, or at least touch only the plain inside of the container, but this is not true of many sweets, chocolates and candies. An example of the dangers resulting from this practice is given in the report referred to above. A coloured trading card, containing 88 mg/kg of lead, was found to be packed in immediate contact with chewing-gum. In addition, in the course of ordinary domestic use, many wrappers, especially of frozen goods packs which are partially used and then returned to the freezer, become torn and come into contact with the contents. This can also result in contamination.

Not all types of paper or plastic used for food contain unacceptably high levels of toxic metals. In many countries regulations control the quality of wrapping materials which may be used on food. In Italy, for instance, a Ministerial Decree[58], sets strict limits for migration into foods of metals and other toxic substances from paper, cardboard, plastic and other packaging materials. Provided such 'food quality' material is used, there is little danger for the consumer. A study carried out by the Canadian Department of Health and Welfare[59] made this clear. X-Ray fluorescence scanning of 62 plastic food containers purchased in local supermarkets showed the presence of barium, bismuth, cadmium, copper, iron, mercury, nickel, lead, selenium, strontium and zinc. In spite of this array of metals shown to be in the plastic, when tested for solubility, no significant amounts of toxic metals could be leached from the containers. Nevertheless, further tests did show that the situation could be changed with prolonged use and exposure to sunlight. Ultraviolet light and surface abrasion increased the solubility of both cadmium and mercury. In addition it was found that if small fragments of some of the plastics were ingested significant amounts of cadmium could be absorbed by the body.

Apart from contamination of food with metals released from inks and dyes used on packaging materials, the use of certain types of colouring substances in the foods themselves have been known to cause problems. Food additive regulations in most countries strictly control the use of such materials in food. However, unapproved colours continue to be used in some countries, and even some approved colours are suspect. An Indian study[60] underlined the problem. Of 36 food colours used commercially, half of which were approved by government regulations and the other half unapproved, all were found to contain arsenic, cadmium, chromium, copper, lead, manganese, nickel and zinc. Though levels of arsenic, chromium, copper and lead in the permitted colours met the legal limits, many of the non-approved colours exceeded them.

CONTAMINATION OF FOOD BY METAL CONTAINERS

Every process to which food is subjected and each type of container in which it is held make some contribution to the load of chemical contaminants which the food can carry into our bodies. Packaging is, in fact, only one source of metal contamination of food among many others[61]. Nevertheless, for historic reasons at least, the possibility of food being contaminated by metals from the container in which it is packed, looms large in public awareness of the problem of food hygiene.

The use of what we know as a 'can' may be attributed to the response of scientists to Napoleon's offer of a prize, in the early nineteenth century, to anyone who could develop a practical method of food preservation which would enable him to keep his vast armies well supplied. One of his subjects, Nicholas Appert, provided the solution, based on his discovery of the principle that spoilage bacteria in food can be destroyed by heat alone. In 1804 he produced preserved meat for the French navy by packing it in glass jars.

Appert's discovery was developed in a practical manner by an Englishman, Peter Durand, who substituted tin-coated iron cans for glass jars. In 1810 he was granted a patent for this technique of packing foods in hermetically sealed metal cans. Napoleon's enemies saw the advantages of canning. Initially cans were produced to meet the needs of the military and of expeditions of discovery. At the Dartford Iron Works, in Kent, John Hall and Bryan Donkin produced canned foods for the British army and navy. Their cans were made of heavy gauge tinplate, laboriously sealed by hand soldering, resulting in a heavy, but remarkably successful food container. Examples of these primitive cans, some made as early as the 1820s for use on a Royal Navy expedition to the Arctic, can still be seen in the British Museum. Subsequently other canned foods became available: sardines (1834); green peas (1837); salmon and tomato (1864); condensed milk (1856). In 1892 the first pineapple cannery was established in Hawaii. The first canned soup was produced in the USA in 1897[62].

Not all aspects of these early attempts at food preservation by canning were successful. It was found that canned foods sometimes contained toxic levels of lead and other metals and the products came to be associated with food poisoning. There was even a move in some countries to ban canning as a means of preserving food. In Germany, a congress of physicians at Heidelberg recommended that 'tin plate should be forbidden for the making of vessels in which articles of food are to be preserved'[63].

Though today the basic principles of canning are the same as they were in the nineteenth century, techniques and the quality of materials used have been greatly improved since those early days. Nevertheless, the problem of metal contamination of the contents of cans has not entirely gone away over the past 180 years. To appreciate why this is we need to look at what canning and tinplate are.

Technically speaking a 'can' is any hermetically sealed container in which food is subjected to a 'canning' process, that is heat treatment to extend shelf life. Though most cans are made from tinplate or aluminium, plastic cans are now beginning to be used in some countries.

Aluminium cans are mainly used for beverages, such as beer and soft drinks, and tinplate cans for food products. Metal contamination of food can result from the use of both types of material.

Tinplate is a composite packaging material made up of a low carbon mild steel base, a very thin layer of iron–tin alloy, a thin layer of pure tin, a very thin layer of oxide and finally a monomolecular layer of 'physiologically safe' edible oil[64]. Each layer has its special role to play. The tin plate is 'passivated', either chemically or electrochemically to build up a 'passive' layer on the surface to prevent oxidation. In both techniques chromium is deposited on the tinplate. The steel base is strong and rigid, giving tinplate its robust characteristics which allow it to withstand high speed canmaking as well as the subsequent hazards of handling and storage. The tin coating gives an almost inert inner surface, while the oxide layer protects the tin from oxidation and corrosion. The outer layer of oil helps to prevent scratching of the can during manufacture and handling.

As was noted with regard to the tin used to plate copper cooking utensils, the high quality or 'pure' tin used in the manufacture of tinplate can contain from 0.2 to 0.25 per cent of other metals. Australian standards, for instance, stipulate a minimum tin content of 99.75 per cent, with a maximum of 0.6 per cent manganese, 0.06 per cent chromium, 0.05 per cent molybdenum and 0.04 per cent nickel in the steel base. These standards are identical to the International Standards Organisation and the EEC specifications[65].

Cans are often coated internally with a lacquer, generally a thermosetting resin polymerised on the surface of the tin during a baking process. This provides additional protection where there is special danger of the food contents corroding the metal. However, in the months or even years during which canned food may be in storage, even this barrier may not be sufficient. A slow corrosion process can take place, especially if the lacquer layer is imperfect. As a result levels of iron, tin and sometimes other metals may build up in the food.

Various factors, such as the amount and type of organic acids, levels of nitrate, amount of oxidising and reducing agents, storage temperatures and the presence or absence of lacquer, determine the amount of corrosion. In addition, until recently, in most cans, lead solder was used to join seams and attach lids. This practice can lead to lead pick-up by the contents of the can. New technologies, using non-lead solder, or even solder-free joins, are being introduced and should eventually overcome this problem. However, as the literature shows, lead contamination of canned food still occurs.

In modern soldered cans, in contrast to those produced not so many years ago, solder comes into direct contact with food in two small areas only; at the top and bottom of the can. Sometimes, however, solder splashes are found elsewhere inside the can. In unlacquered cans it is relatively easy to remove these splashes before the can is filled, but it is not so easy to do with lacquered cans. Lead-containing particles can remain embedded in the lacquer layer and thus come into contact with food.

It is of interest, as well as an illustration of the electrochemical principles of the cathodic behaviour of lead relative to tin (i.e. when both metals are in electrical contact, the tin will dissolve in preference to the lead and lead already in solution will be replaced by the tin), that analysis of the contents of a tin of roast veal which had been taken on an Arctic expedition in 1837 and left there in the permanent deepfreeze until it was recovered a century later, contained 71 mg/kg of iron, 783 mg/kg of tin and only 3 mg/kg of lead. A 100-year-old tin of carrots recovered from the same Arctic cache had 308 mg/kg of iron, 2440 mg/kg of tin and no lead in its contents. These are remarkable results, especially when it is realised that these were primitive tinned containers, in which lead solder was used extensively to attach the tops and side seams. Unfortunately there is some evidence that not all of the food canned in this way was safe when eaten closer than 100 years to the date of manufacture. The permafrost-preserved bodies of members of the disastrous Franklin Expedition of 1846, which had set out to find the North-West Passage, showed evidence of lead poisoning when autopsied. It is probable that the lead came from the stock of 8000 cans of food on which they had lived during a winter trapped in the ice.

For most canned foods there is a limit to storage time beyond which they are no longer fit for consumption. An investigation of canned foods on sale in Israel[66] found that many products had been on shop shelves for more than three years. A can of fruit juice had more than 700 mg/litre of tin and 2 mg/litre of lead. The oldest cans had the highest levels of metal contaminants. An extensive study of canned foods in Italy confirmed this finding[67]. In some foods which had been stored for one year, lead had reached levels of 20 mg/kg. Copper, iron, zinc and manganese also increased in concentration in the food during that time. Levels of zinc, for example, increased by 32 per cent in 208 days and by 58 per cent in one year. It was concluded that migration of iron, zinc and lead in a can of mixed vegetable salad in vinegar rendered the product unfit for consumption after 270 days.

The effect of lacquering on metal migration in canned products has been illustrated in a number of studies. Investigators in New Zealand looked at lead in foods, including baby foods, contained in lacquered and unlacquered cans[68], and concluded that lead contamination is mainly associated with lacquered cans. However, their results showed considerable inter-can variation, for lead as well as for other metals. Blackcurrants in a lacquered can contained 10, 160 and 2600 mg/kg of lead, tin and iron respectively, while rhubarb, also in a lacquered can, had 0.3, 30 and 24 mg/kg of the same metals. In contrast, gooseberries in a plain can had 0.3, 260 and 4.3 mg/kg, and pineapple, also in a plain can, 0.35, 105 and 3.0 mg/kg of the same metals. These variations may have been related to the pH of the relatively acidic foods analysed. Other foods, of a less acidic nature, had the following metal content: green beans, in a lacquered can, 0.7, 10 and 4.8 mg/kg, and broad beans, in an unlacquered can, 0.02, 10 and 18 mg/kg of lead, tin and iron.

Storage temperature may also have an influence on metal uptake from the can by food, as has been shown in a study by Catala and Duran on levels of tin in canned green beans stored at different temperatures[69]. They found that while tinplate behaved satisfactorily at 20° C, at 37° C it could be completely de-tinned over a two-year period, with the tin content rising to over 100 mg/litre in the liquid and 1000 mg/kg in the beans.

Opening a can of food and leaving its contents exposed to the air can also increase the amount of tin entering canned food from the plate.

The level of nitrate in a canned food is also related to tin uptake[70]. It has been suggested, moreover, that the presence of nitrate increases the toxicity of the metal. An outbreak of food poisoning in the Saar, Germany, was traced to the consumption of canned peaches imported from Italy. The fruit contained about 400 mg/kg of tin, compared to the general level of 44–87 mg/kg in canned peaches produced elsewhere. In addition the Italian cans contained a high level of nitrate. This was traced to the use of well water which had upwards of 300 mg/litre of NO_3. It was concluded by the investigators that the level of tin in the peaches was not in itself likely to be toxic, but in combination with the nitrate, was responsible for the illness.

As has been mentioned above, the tin used in plating is not 100 per cent pure. In the UK and several other countries it may contain 0.2 per cent of other metals. Elsewhere different standards apply. Polish legislation, for example, specifies 99.75 per cent purity of tin used in plating. The non-tin fraction may contain up to 0.05 per cent arsenic, in addition to lead. It is estimated that use of such plate could result in between 0.244 and 0.380 mg/kg of arsenic entering food, depending on the size of the can.

Modern developments in canning technology may soon largely eliminate some of the problems of metal contamination, especially by lead, of foods preserved in this way. Use of the welded can is rapidly growing, thus eliminating lead solders and the possibility of contamination from this source[71]. In the Federal Republic of Germany all major food can manufacturers produce lead-free containers. This is the normal practice in Australia and most other technologically-advanced countries[72].

Aluminium Containers

The aluminium can, for soft and brewed alcoholic drinks, is a feature of modern life. We make enormous use today of aluminium foil for food wrapping and for the manufacture of a variety of food containers. This is a relatively recent development, for though aluminium has been commercially available for more than half a century, technological difficulties prevented its widespread use in the food industry for many years. Initially, too, there was some opposition, on the grounds of health, to its use in connection with food. However, this opposition was overcome, and today it is probably the most widely used metal for food packaging. As we shall see in a later chapter, in recent years there has been a re-emergence of health concerns regarding aluminium in food, though it is still accepted by the majority of health authorities that use of the metal in food packaging and processing is perfectly safe.

As with other packaging metals, the pure metal is not used commercially. To provide strength, improve formability and increase corrosion resistance, various alloying elements are added to the aluminium. These include iron, copper, zinc, manganese and chromium. These metals, as well as aluminium itself, may migrate into the contents of a container if corrosion takes place. Uncoated aluminium cans are readily attacked by food acids, causing hydrogen release and can swell. Aluminium has been found to be an excellent metal for storage of both alcoholic and non-alcoholic beverages. It has been shown that even when solution of the metal occurs, the flavour will not be affected until more than 10 mg/litre of aluminium is dissolved. It is claimed that beers packed in aluminium cans are superior in colour, flavour and clarity to those packed in tin-plate[73].

Aluminium acts as a 'sacrificial anode' when used in conjunction with tinplate or steel in a can. The use of aluminium ends on such cans takes advantage of this characteristic to provide electrochemical protection for the contents of the can.

In spite of these useful properties of aluminium, it has been found that it is advantageous to coat the inside of aluminium cans with a vinyl epoxy or some other resin. This prolongs the shelf life of canned alcoholic beverages and soft drinks. Strong alcoholic drinks, such as whisky and brandy and some wines, can cause undesirable reactions with aluminium if packed in unlacquered cans. Pitting corrosion can occur as well as discolouration of the beverage. As aluminium is dissolved by the liquid, a flocculent precipitate of aluminium hydroxide can also form.

Aluminium foil is extensively used for food wrapping and packaging as well as during cooking. If properly handled and used in the right conditions, the foil fulfils these roles excellently. However, in the case of frozen foods, for example, bad handling and, especially, poor storage and resulting thawing, can cause both deterioration of the food and corrosion of the aluminium container. In addition, if foil comes into contact with other metals, an electrochemical reaction can take place, with the aluminium acting as a sacrificial anode. Pitting and corrosion of the aluminium foil will occur and the quality of the food may be affected.

CONTAMINATION OF FOOD BY RADIOACTIVE METALS

Eleven years ago, when the first edition of this book was published, the consideration of radioactive contamination of food was largely a theoretical exercise, except in the most unlikely situations. Events over that decade, such as the Three Mile Island reactor leak, the build up of radioactive waste in the Irish Sea and, especially, the Chernobyl accident, have, unfortunately, made such contamination a much more practical problem for food manufacturers and consumers. When shipments of meat and other foods in international trade are blocked at ports of entry by quarantine officers, and farmers in peaceful parts of Britain are forbidden to sell their products because of radioactive contamination in farm animals, then what was so recently a remote possibility must be recognised as a real threat.

It was just over a hundred years ago, in 1896, that Becquerel discovered that the element uranium emitted rays which could affect the emulsion on a photographic plate. A decade and a half later Madame Curie and her husband Pierre showed that atoms of uranium and radium undergo spontaneous disintegration to form atoms of other elements and in the process emit penetrating rays. This spontaneous decomposition of atomic nuclei is what we call radioactivity.

Several radioactive elements and isotopes of elements occur in nature and man has been exposed to their penetrating rays to a greater or lesser extent ever since he first appeared on earth. But during the past five decades a new form of radioactivity, artificially produced, has been introduced into the environment. Atomic weapons and nuclear energy production have added greatly to the earth's radioactive load and, as a consequence, contributed to the problem of food contamination.

The penetrating rays which are emitted by radioactive elements are commonly referred to as ionising radiation because of the effects they produce. The rays are high-velocity particles emitted from the nuclei of atoms. Two principal kinds of particles are produced: α-particles, which consist of two neutrons and two protons (identical to the helium nucleus); and β-particles, which are high-velocity electrons. These emissions may be accompanied by γ-rays. These are electromagnetic rays, like X-rays, and are more penetrating than either the α- or the β-particles.

When a nucleus loses an α-particle, its atomic mass is decreased by four units and its atomic number by two. The loss of two protons from the nucleus is accompanied by the loss of two electrons, thus maintaining electrical balance. When a β-particle is lost, there is no change in the mass of the nucleus, but there is an increase of one in its net positive charge, so that the atomic number increases by one unit. The emission of γ-rays has no effect either on the atomic number or the atomic weight of an element, since these rays possess neither mass nor charge.

All elements with atomic numbers greater than that of bismuth have one or more isotopes which are naturally radioactive. A few elements of lower atomic number, such as potassium and rubidium, also have naturally occurring radioactive isotopes. Artificially produced radio-isotopes of many other atoms are also known.

The terminology used with regard to isotopes and radioactive elements needs some clarification. 'Isotope' refers to a type of atom, while 'nuclide' is used to refer to the nucleus of a specific isotope characterised by its atomic number, mass number and its energy. Radionuclides are unstable nuclides which emit ionising radiation.

The rates at which disintegration of radionuclides take place are not the same for all elements. The term 'half-life' refers to the time taken for the activity of a radionuclide to decay to half its original level, that is, for half the nuclei to disintegrate. The SI (Systeme International d'Unites) unit used for radioactivity is the becquerel (Bq) which is defined as one radioactive disintegration per second.

Radioisotopes in Food

Three different types of natural radionuclides can occur in food. Some, such as uranium and thorium, have always been with us. They have extremely long half-lives of millions of years. A second group is produced as 'daughter-elements' by the disintegration of the first group. Radium 226, produced by uranium 238, is one of these. Radium, in turn, has a daughter of its own, radon, which is, again, unstable, producing its own daughter radionuclides, lead 210 and polonium 210.

Another group of radionuclides is constantly being formed by the action of cosmic rays on elements in the atmosphere. One important product of this action is carbon 14, made from the nucleus of nitrogen.

These naturally occurring radionuclides account for the 'natural background radiation' in the presence of which we have evolved[74]. Unfortunately, man-made additions to that background over recent decades have more than doubled the level of radiation to which we are exposed. Our exposure to radiation has greatly increased since nuclear weapons were developed in the mid 1940s. The increase was especially rapid in the 1960s when the nuclear powers carried out many atmospheric tests. The development of the nuclear energy industry, as well as of applications of radioactivity to medicine, have also added to the world's radioactive background.

When an atom bomb explodes, or a nuclear reactor operates, a complex mixture of many different radionuclides is produced by fission of the fuel materials. Included are the very long half-lived heavy elements such as uranium 235 and plutonium 239, as well as caesium-134 (half-life 30 years), strontium 90 (half-life 28 years), cobalt 60 (half-life 5.27 years), ruthenium (half-life 1 year) and the non-metal iodine 131 (half-life 8 days). Many other radionuclides are also produced, not directly from the bomb's fuel, but as a result of action of the neutrons from the fission on other elements present in the casing of the bomb or in the immediate environment of the explosion. The 'activation products' include zinc 65 (half-life 245 days) and the non-metal carbon 14 (half-life 5760 days). The radioisotopes used in medicine, industry and research are also activation products, but made by 'neutron bombardment' of selected target atoms in specially designed nuclear reactors.

Radioactive Contamination of Foods

Man-made radionuclides from a variety of sources (testing of nuclear weapons, release of waste into air and water from nuclear plants, accidents in nuclear power stations and in nuclear submarines and satellites), can all result in contamination of foodstuffs with radionuclides.

Radionuclides released into the atmosphere will eventually fall to earth where they can be absorbed by plants and be consumed by grazing animals. Radionuclides discharged into fresh or marine waters can result in contamination of fish and other aquatic organisms. Through these pathways they can eventually enter the human food chain and bring with them hazards to health.

The consequences of such contamination and the precautions which need to be taken in order to keep such consequences to a minimum, have been summarised by O'Flaherty in a paper presented at a conference on pure food production held in Dublin in 1988 when many people in that part of the world were much concerned, not only by the recent Chernobyl accident but also by the growing level of radioactivity in the Irish Sea caused by discharges from the Sellafield nuclear fuel reprocessing plant in Cumbria[75].

The author pointed out that it is necessary to consider the relationship between a given level of radioactivity in a particular food item and the radiation dose to an individual which would result from eating that food. He noted that for any given food, the dose depends on four factors:

1. the quantity of the food eaten over a period of time;
2. the level of radioactivity in the food consumed, measured in Bq/kg or Bq/litre;
3. the type of radiation (alpha, beta or gamma) emitted by the radionuclide(s) in the food;
4. the chemical properties of the radionuclides which determine whether they are uniformly distributed throughout the body or concentrated in an organ, like iodine in the thyroid gland.

The effects of these factors are summarised in the concept of a 'Dose Conversion Factor', which relates to the amount of radioactivity consumed (in Bq) to the resulting 'Dose Equivalent'. This is expressed in SI units as sieverts (Sv), and is the quantity obtained by multiplying the absorbed dose by a factor which allows for the different effectiveness of the different types of ionising radiation in harming tissues. This is usually taken as 1 for beta and gamma radiation and 20 for alpha particles.

In order to determine the radiation dose which an individual will receive from a particular contaminated food, it is necessary to know the level of radioactivity of each radionuclide in the food as well as the actual amount of that food consumed. Dietary surveys can provide information on the latter. The identity and amount of radionuclides present in the foods are determined by use of a high-resolution gamma spectrometer. These may be the gamma-emitting caesium isotopes, Cs-134 and Cs-137, and iodine 131 (I–131). Other procedures are used to detect the generally much lower levels of alpha-emitting radionuclides such as plutonium 239 (Pu-239) and of beta-emitting radionuclides such as strontium 90 (Sr-90).

The detrimental effects of radiation on health are either non-stochastic, in which the severity of the effect is related to the magnitude of the dose received, or stochastic, in which the probability of an effect occurring, but not its severity, is related to the dose received. Non-stochastic effects, which include various degrees of radiation sickness such as were seen in the firefighters of Chernobyl, occur only at relatively large doses. Stochastic effects may be a cancer or a genetic effect, which is not immediate and for which there may be no threshold dose below which it does not occur. In radiological protection it is assumed that no dose threshold exists and therefore every dose, no matter how small, is considered to carry a risk.

It is a matter of major international concern that effective measures should be applied to protect the community from exposure to unnecessary radiation doses, whether from radioactivity in food or from any other cause. It is desirable that this should be done without having to impose measures so restrictive as to cause hardship or economic damage altogether disproportionate to the level of risk. The International Commission for Radiological Protection[76] has set dose limits appropriate to different circumstances which are believed to offer reasonable protection. Of course, the reasonableness of the limits is questioned by those who maintain that there can never be a safe level of radiation.

Protective measures recommended will depend on the circumstances. Initially it may be necessary to prohibit grazing by animals or to restrict sale for human consumption of milk from particular areas. The use of vegetables grown in contaminated soils, or the importation of foodstuffs from certain countries, may also be restricted. The overall aim will be to limit the sale and movement of food with radioactive contamination above an agreed level. What those particular intervention levels are will depend on which national or international guidelines are followed. There are, in fact, considerable differences between practices in different countries.

Intervention levels established by the World Health Organization, for instance, are much less restrictive than those promulgated by the Food and Agriculture Organization of the United Nations. Both organisations based their levels on different sets of criteria. A third set of intervention levels has been established by the European Community. The EC standards are contained in two regulations, the first of which prescribes limits on Cs-134 and Cs-137 of 370 Bq/kg in milk and infant foods and 600 Bq/kg for all other foods imported from other countries. The second lays down maximum permitted levels of contamination by a range of radionuclides in different foods. The levels lie about mid way between the WHO and the FAO intervention levels. Unfortunately, different countries have adopted different guidelines, thus creating a lack of harmony which affects international trade in foodstuffs. Different 'action levels' applied in different countries are tabulated in a a paper by Cope on the international implications of the Chernobyl accident[77].

The reality of radioactive contamination of food was brought home to Europeans in particular by the Chernobyl accident. Before that concern had been caused by reports of lesser accidents at nuclear power stations and processing plants in various parts of the world, and of environmental contamination by radioactive waste, sometimes released deliberately. However, on 26 April 1986, there occurred at Chernobyl, some 100 km north of Kiev in the Ukraine, what has been described as 'the worst accident in the history of commercial nuclear power generation'[78]. A reactor at the plant accelerated from a fraction of full power to 100 times full power in 4 seconds, causing an explosive generation of steam. The explosion lifted the enormous 1000-tonne roof of the building and spewed burning debris over the surrounding area. The reactor core was rich in a wide range of fission products, many of them highly active elements among which iodine 131, caesium 137 and tellurium 132 were the most prominent. Extensive contamination of land around the plant occurred, requiring the evacuation of more than 135000 people. A radioactive cloud was produced and was fed by emissions from the damaged plant over several days. It was carried in a series of atmospheric plumes over all parts of Europe, as well as east towards Japan and China. Heavy rain caused deposition of large amounts of radioactivity in parts of Scandinavia and elsewhere in Europe as the plumes moved on. One of the plumes crossed the English Channel on 2 May, 6 days after the accident, and after travelling more than 4000 km on a tortuous track from the Ukraine. Within two days the plume had moved across England, Wales and Scotland and had spread over the Irish Sea to Ireland. Considerable public concern was caused by the passage of the cloud.

Monitoring by health authorities and others of drinking water, milk, vegetables and meat in various parts of Europe including the UK and Ireland, found levels approaching, or above, intervention levels in several types of food. A variety of restrictions were imposed on the sale or use of certain foodstuffs. Supplies of vegetables, particularly spinach, were withdrawn from sale in France, Greece, Italy, Luxembourg, the Netherlands and West Germany. Reindeer meat sales were restricted in Lapland. In the UK restrictions were imposed on the slaughter of lambs grazing on contaminated grass in parts of Wales, Cumbria and Scotland.

The Chernobyl accident brought home to the international community the risk involved in operating nuclear installations and the need for high standards of construction and safety. Unfortunately, as was noted by Cope in the Report already referred to, it also revealed the European Community's inability to respond in a coordinated way to a major disaster. Nevertheless, the accident provided a stimulus for the introduction in many countries of improved monitoring of radioactive contamination of foods.

Concern about levels of radionuclides in the diet, and the implications for health of radioactive emissions from the nuclear industry, did not, of course, originate with Chernobyl. Indeed, as data reported by O'Flaherty[75] show, the Russian accident, while locally very serious, caused a lower level of contamination of food in the UK and Ireland than had been caused by some local activities and accidental discharges. Table 9, taken from his paper, indicates that levels of radioactive caesium in fish in the Irish Sea, for instance, probably owes more to discharge from the Sellafield reprocessing plant in Cumbria than to fallout from Chernobyl.

O'Flaherty also notes that overall, even after Chernobyl, dose levels of radioactivity from all sources in Ireland are 'relatively small, being at most of the order of 10–15 percent of the ICRP dose limit of 1 mSv for members of the public. This also bears comparison with the annual radiation dose of about 2 mSv, which is received by every member of the population from normal causes'. However, O'Flaherty's final comment is also true and might well be extended to countries other than his own: 'small though

TABLE 9

Mean Concentrations of Total Caesium (Cs-134 + Cs-137) in Fish and Shellfish from the Irish Sea and Resulting Radiation Dose to Heavy Fish Eaters (200 g/day fish; 20 g/day shellfish) 1982–86 in Ireland

Year	Total Cs (Bq/kg)		Radiation dose (μSv)
	Fish	Shellfish	
1982	66	31	71
1983	53	30	58
1984	45	20	49
1985	28	9	25
1986	21	8	20

these doses are, the contamination of food by radionuclides is naturally and properly a matter of serious concern both to the Irish consumer and to other countries who buy our food exports. It is therefore of vital importance that the systematic and comprehensive programme of monitoring of radioactivity in food, which is conducted by the Nuclear Energy Board, should be continued for the foreseeable future'.

REFERENCES

1. DITTMER, N.J. (1937). A quantitative study of the roots and root hairs of a winter rye plant. *Am. J. Bot.*, 24, 417–20, as quoted in E. Epstein, (1972). *Mineral Nutrition of Plants* (J.Wiley, New York).
2. HALL, N.S., CHANDLER, W.F., VAN BAVEL, C.H.M. *et al.* (1953). A tracer technique to measure growth and activity of plant root systems. *North Carolina Agric. Exp. Sta. Tech. Bull.*, 101, 1–40.
3. LISK, D.J. (1972). Trace metals in soils, plants and animals. *Adv. Agron.*, 24, 267–320.
4. GARDINER, M.R., ARMSTRONG, J., FELS, H. and GLENROSS, R.N. (1962). A preliminary report on selenium and animal health in Western Australia. *Aust. J. Expt. Ad. Anim. Husb.*, 2, 261–9.
5. REILLY, C. (1969). The uptake and accumulation of copper by *Becium homblei*. *New Phytol.*, 68, 1081–7.
6. DUVIGNEAUD, P. and DENAEYER-DE SMET, S. (1963). La vegetation du Katanga et des sols metalliferes. *Bull. Soc. Roy. Bot. Belg.*, 96, 93–231.
7. REILLY, A. and REILLY, C. (1973). Zinc, lead and copper tolerance in the grass *Stereochlaena cameronii*. *New Phytol.*, 72, 1041–6.
8. BRADSHAW, A.D., McNEILLY, T. and GREGORY, R.P.G. (1965). Industrialisation, evolution and the development of heavy metal tolerance in

plants, in *Ecology and Industrial Society*, ed. Goodman, G.T., Edwards, R.W. and Lambert, J.M., *Br. Ecol. Soc. Symp.*, 5, 327–34.

9. REILLY, C. (1972). Amino acids and amino acid-copper complexes in water-soluble extracts of copper-tolerant and non-tolerant *Becium homblei*. *Z. Pflanzenphysiol.*, 66, 294–6.

10. MORGAN, H. (ed.) (1988). *The Shipham Report. An Investigation into Cadmium Contamination and its Implications for Human Health.* (Elsevier, London).

11. REILLY, A. and REILLY, C. (1972). Patterns of lead pollution in the Zambian environment. *Med. J. Zambia*, 6, 125–7.

12. HANSEN, J.C., WULF, H.C., KROMANN, N. and ALBGE, K. (1983). Mercury levels in the diet and blood of Greenlanders. *Sci. Total Environ.*, 26, 233–43.

13. REILLY, C., SUDARMADJI, S. and SUGIHARTO, E. (1989). Mercury and arsenic in soil, water and foods of the Dieng Plateau, Java. *Proc. Nutr. Soc. Aust.*, 14, 118.

14. PENNINGTON, J.A.T. (1988). Aluminium content of food and diets. *Food Add. Contam.*, 5, 161–232.

15. WENLOCK, R.W., BUSS, D.H. and DIXON, E.J. (1979). Trace nutrients 2. Manganese in the British diet. *Br. J. Nutr.*, 41, 253–61.

16. BRISBANE CITY COUNCIL (1989). *Report on Domestic Refuse Tips* (Brisbane, Queensland, Australia).

17. MACKENZIE, E.J. and PURVES, D. (1975). Toxic metals in sewage sludge. *Chem. Ind.*, 4 January, 12–13.

18. HARRISON, R.M. (1978). Metals in household dust. *Sci. Total Environ.*, 8, 89–97.

19. PRICE, B. (1988). Cleaner solutions from rubbish tips. *New Scientist*, 114, 47–50.

20. BERROW, M.L. and WEBBER, J. (1972). The use of sewage sludge in agriculture. *J. Sci. Food Agric.*, 23, 93–100.

21. PIKE, E.R., GRAHAM, L.C. and FOGDEN, M.W. (1975). Metals in crops grown on sewage-enriched soil. *JAPA*, 13, 19–33.

22. CAPON, C.J. (1981). Mercury and selenium content and chemical form in vegetable crops grown on sludge-amended soil. *Arch. Environ. Contam. Toxicol.*, 10, 673–89.

23. STRENSTROM, T. and VAHTER, M. (1974). Heavy metals in sewage sludge for use on agricultural soil. *Ambio*, 3, 91–2.

24. WILLIAMS, C.H. and DAVID, D.J. (1973). Heavy metals in Australian agricultural soils. *Aust. J. Soil Res.*, 11, 43–50.

25. KJELLSTRÖM T., LIND, B., LINNMAN, L. *et al.* (1975). Variations of cadmium content of Swedish wheat and barley. *Arch. Environ. Health.*, 30, 321–8.

26. ALFTHAN, G. (1988). Longitudinal study on the selenium status of healthy adults in Finland during 1975–1984. *Nutr. Res.*, 8, 467–76.

27. BAKIR, F., DAMLUJI, S.F., AMIN-ZAKI, L. *et al.* (1973). Methylmercury poisoning in Iraq. *Science*, 181, 230–41.

28. MINISTRY OF AGRICULTURE, FISHERIES AND FOOD (1982). *Survey of Arsenic in Food: the 8th Report of the Steering Group on Food Surveillance*;

the *Working Party on the Monitoring of Foodstuffs for Heavy Metals* (HMSO, London).
29. ASSOCIATION OF PUBLIC ANALYSTS. (1971). *Joint Survey of Pesticide Residues in Foodstuffs sold in England and Wales.*
30. JERVIS, R.E. (1970). *Mercury Residues in Canadian Foods, Fish and Wildlife* (University of Toronto Press).
31. PHELPS, R.W., CLARKSON, T.W., KERSHAW, T.G. and WHEATLEY, B. (1980). Methyl mercury contamination of fish in a Canadian lake. *Arch. Environ. Health*, 35, 161–8.
32. PRATER, B.E. (1975). Water pollution in the river Tees. *Water Pollut. Control*, 74, 63–76.
33. WORKING PARTY ON MATERIALS CONSERVATION AND EFFLU-ENTS (1975). *Report of the Industrial and Technical Committee of the Institute of Metal Finishing, Trans. Inst. Metal Finish.*, 53, 197–202.
34. BRENNAN, J.G. (1976). *Food Engineering Operations*, 2nd edn (Elsevier Applied Science Publishers, London).
35. GILFILLAN, S.C. (1965). Lead poisoning in ancient Rome. *J. Occup. Med.*, 7, 53–60.
36. BERITIC, T. and STAHULJAK, D. (1961). Lead poisoning from pottery vessels. *Lancet*, 1, 669.
37. BEECH, F.W. and CARR, K.G. (1977). Cider and perry, in *Economic Microbiology*, Vol. 1. *Alcoholic Beverages*, ed. Rose, A.H., 130–220 (Academic Press, London and New York).
38. REILLY, C. (1989). Unpublished observations.
39. HARPER, W.J. and HALL, C.W. (1976). *Dairy Technology and Engineering* (Avi, Westport, Conn.).
40. WHITMAN, W.E. (1978). Interactions between structural materials in food plant, and foodstuffs and cleaning agents. *Food Progress*, 2, 1–2.
41. SAMPAOLO, A. (1971). Migration of metals from stainless steel. *Rass. Chim.*, 23, 226–33.
42. SCHOCK, M.R. and NEFF, C.H. (1988). Trace metal contamination from brass fittings. *J. Am. Water Works Assoc.*, 80, 47–56.
43. REILLY, C. (1978). Copper and lead uptake by food prepared in tinned-copper utensils. *J. Food Technol.*, 13, 71–6.
44. REILLY, C. (1976). Contamination of food by lead during catering operations. *Hotel Catering and Institutional Management Rev.*, 2, 34–40.
45. MACPHAIL, A.P., SIMON, M.O., TORRANCE, J.D., CHARLTON, R.W., BOTHWELL, T.H. and ISAACSON, C. (1979). Changing patterns of dietary iron overload in black South Africans. *Am. J. Clin. Nutr.*, 32, 1272–8.
46. DROVER, D.P. and MADDOCKS, I. (1975). Iron content of native foods. *PNG Med. J.*, 18, 15–17.
47. TANNER, M.S., BHAVE, S.A., KANTARJIAN, A.H. and PANDIT, A.N. (1983). Early introduction of copper-contaminated animal milk feeds as a possible cause of Indian childhood cirrhosis. *Lancet*, ii, 992–5.
48. PENNINGTON, J.A.T. (1987). Aluminium content of foods and diets. *Food Add. Contam.*, 5, 161–232.
49. REILLY, C. (1985). The dietary significance of adventitious iron, zinc, copper and lead in domestically prepared food. *Food Add. Contam.*, 2, 209–15.

50. SMART, G.A. and SHERLOCK, J.C. (1985). Chromium in foods and the diet. *Food Add. Contam.*, 2, 139–47.
51. KLEIN, M., NAMER, R., HARPUR, E. and CORBIN, R. (1970). Earthenware containers as a source of fatal lead poisoning. *New Eng. J. Med.*, 283, 669–72.
52. MINISTRY OF AGRICULTURE, FISHERIES AND FOOD (1974). *Report of the Inter-Departmental Working Group on Heavy Metals* (HMSO, London).
53. EUROPEAN ECONOMIC COMMUNITY (1984). *Council Directive No. 84/500, Off. J. E.C.*, 27, 12–16.
54. FINNISH MINISTRY OF HEALTH (1962). *General Regulation for Foodstuffs, Decree No 408.* (Helsinki).
55. INTERNATIONAL ORGANISATION FOR STANDARDISATION (1981) *International Standard*, ISO 6486/1,2.
56. WATANABE, Y. (1974). Cadmium and lead on decorated glass drinking glasses. *Ann. Rep. Tokyo Metropol. Lab. Pub. Health*, 25, 293–6.
57. HEICHEL, G.H., HANKIN, L. and BOTSFORD, R.A. (1974). Lead in coloured wrapping paper. *J. Milk Food Technol.*, 37, 499–503.
58. GRAMICCIONI, L. (1984). Migration of metals into food from containers. *Rass. Chim.*, 36, 271–3.
59. MERANGER, J.C., CUNNINGHAM, H.M. and GIROUX, A. (1974). Plastics in contact with foods. *Can. J. Pub. Health*, 65, 292–6.
60. KHANNA, S.K., SINGH, G.B. and HASAN, M.Z. (1976). Toxic metals in food colourings. *J. Sci. Food Agric.*, 27, 170–4.
61. SELBY, J.W. (1967). *Food Packaging, the Unintentional Additive*, in *Chemical Additives in Food*, ed. Goodwin, R.W.L., 83–94 (Churchill, London).
62. FOOD AND AGRICULTURAL ORGANISATION (1986). *Guidelines for Can Manufacturers and Food Canners. Introduction*, p.1 (FAO, Rome).
63. VAN HAMEL ROOS (1891). On tin in preserved articles of food. *Rev. Intn. Falsif.*, 4, 179 (abstract in *Analyst* (1891), 16, 195).
64. CANNED FOOD INFORMATION SERVICE (1988). *Food Canning: An Introduction* (Food Canning Information Service, Melbourne, Australia).
65. BOARD, P.W. and VIGNANOLI, L. (1976). Electrolytic tinplate for some Australian canned foods. *Food Technol. Aust.*, 28, 486–7.
66. BECKHAM, I., BLANCHE, W. and STORACH, S. (1974). Metals in canned foods in Israel. *Var Foeda*, 26, 26–32.
67. BRANCA, P. (1982). Uptake of metal by canned food with length of storage. *Bull. Chim. Lab. Prov.*, 33, 495–506.
68. PAGE, G.G., HUGHES, J.T. and WILSON, P.T. (1974). Lead contamination of food in lacquered and unlacquered cans. *Food Technol. N.Z.*, 9, 32–5.
69. CATALA, R. and DURAN, I. (1972). Effect of temperature on levels of tin in canned vegetables. *Rev. Agroquim. Technol. Aliment.*, 12, 319–28.
70. BOARD, P.W. (1973). The chemistry of nitrate-induced corrosion of tinplate. *Food Technol. Aust.*, 25, 15–16.
71. LARKEN, D. (1983). The welded can. *Food Technol. Aust.*, 35, 384–7.
72. EGGER, P. (1983). Superwima welding. *Food Technol. Aust.*, 35, 383.
73. JIMINEZ, M.A. and KANE, E.H. (1974). *Compatibility of Aluminium for Food Packaging* (Am. Chem. Soc., Washington, DC).
74. EHRLICH, P.R., EHRLICH, A.H. and HOLDEN, J.P. (1973). *Human Ecology. Problems and Solutions*, 139 (Freeman, San Francisco).

75. O'FLAHERTY, T. (1988). Reducing the risk to human health from radioactivity in the food chain, in *Proc. Conf. Pure Food Production: Implications of Residues and Contaminants.*, 61–78 (The Agricultural Institute, Dublin).
76. ICRP PUBLICATION NO. 40. (1984). Protection of the public in the event of major radiation accidents: Principles for planning (Pergamon Press, for ICRP, Paris).
77. COPE, D. (1988). International dimensions of the implications of the Chernobyl accident for the United Kingdom, in *The Chernobyl Accident and its Implications for the United Kingdom*, eds Worley, N. and Lewins, J., 71–81 (Elsevier Applied Science, London).
78. CENTRAL ELECTRICITY GENERATING BOARD (UK). Report, May 1987 (CEGB, Barnwood, Gloucester).

CHAPTER 3

PROTECTING THE CONSUMER: FOOD LEGISLATION

ADULTERATION OF FOOD

'No person shall sell to the prejudice of the purchaser any article of food or any thing which is not of the nature, substance or quality demanded by such purchaser'. With these words the *Sale of Food and Drugs Act* 1875, the first effective Act of Parliament in the English-speaking world relating to food quality, was passed at Westminster. It laid the foundation for modern food laws in Britain and in those countries that have inherited her legal system. These very words, 'nature, substance, quality' are enshrined in many modern food laws, including the Australian Food Legislation of 1986[1].

Under the 1875 Law sellers of food were to be for ever more obliged to see to it that their goods conformed to the requirements of the Law. In other countries, where the legal system owes little, if anything, to Britain, similar laws and controls were also enacted about the same time to protect the rights of purchasers. Those laws have subsequently been modified in many ways, their details clarified and amplified in response to new developments in food supply and technology. International organisations, principally the United Nations and the European Economic Community, have adopted food laws based on the same philosophy which protects the rights of the purchaser of food and are binding on all member states.

Consumers have not always been so well protected. There is ample evidence to show that legislation to control levels of contaminants in food was sorely needed in pre-1875 days. Accum's classical *Treatise on Adulterations of Food and Culinary Poisons*, published in London in 1820[2] is often quoted in this regard, but there are many other examples in the scientific and even popular literature of the time.

66

Before the days of refined analytical procedures, a problem of major concern was the deliberate contamination of foodstuffs with metals or their salts. This adulteration was practised for one of two reasons: either inert material was incorporated into a product in order to increase its bulk and hence the trader's profits, or, secondly, fraudulent additions of substances were made to improve the appearance and attract the customer's eye[3]. Chalk (natural calcium carbonate) and alum (naturally occurring crystalline potassium aluminium sulphate) were mixed with flour as a cheap 'extender'; copper sulphate crystals were put into beer and 'Prussian Blue' (potassium ferric ferrocyanide) into tea, to improve colour. Other metallic adulterants were used even well into the latter half of the nineteenth century, in spite of the passing of the Food Act: talc and French chalk (both natural forms of magnesium silicate) as well as calcium sulphate, lead chromate and iron oxide were all known to shady food sellers.

Early volumes of *The Analyst*, a 'monthly journal devoted to the advancement of analytical chemistry' (to quote its editorial policy) which was begun, significantly, in 1875, the year of the passing of the Act, by the Society of Public Analysts, make interesting reading in this regard. They give many examples of fraudulent, and sometimes merely ignorant, practices of food processors and sellers, and indicate that the skills of Public Analysts were in demand in ensuring that the provisions of the Food Act were respected. The journal's policy of carrying reports from other countries besides Britain is of considerable interest since it shows that Britain's Continental neighbours also had their problems with food contamination and adulteration. In 1891, for instance, a paper by J. Mayrhofer of the Bavarian Chemical Society on the use of copper salts to preserve the green colour of cooked vegetables, was reprinted from the German *Chemische Zeitung*[4]. Several other papers on this topic were published in *The Analyst* about the same time. One of these noted that there had been a number of prosecutions for the offence of selling green peas to which copper salts had been added[5].

Vegetables were not the only foodstuff to be adulterated with metallic salts. Cheese was also abused in this way. A report on a meeting of the Society of Public Analysts in 1897 recorded that the members had heard that metallic lead had been found in a sample of Canadian cheese and that in Britain a mixture known as 'cheese spice' was used by some manufacturers to prevent heaving and cracking of the product. The 'spice' contained 48 per cent zinc sulphate. The report further added that 'it is well known that the green mould in certain kinds of cheese has been imitated by the insertion of copper or brass skewers'. Some even more serious cases of cheese adulteration were noted: 'instances have occurred in which preparations of arsenic have been added to cheese as a preservative, and in 1841 several persons were poisoned by this means. A similar case occurred in 1854 when a Parisian family suffered, but not fatally'[6]. Today's problem of relatively high levels of aluminium in some processed cheeses seems insignificant when compared with these nineteenth century cases of adulteration.

It is of interest that analysts of a hundred years ago were aware of a problem which is still with us today. An article from a continental journal abstracted in *The Analyst* noted that the use of coloured wrapping materials, such as yellow cheese cloth and the papers on 'chocolate, bonbons etc', contained lead chromate[7]. Salts were illegally added to spices according to another abstract. Cayenne pepper, for instance, was adulterated with 'barium-ponceau lake' to give it 'a brilliant fiery colour'[8].

These few examples give an idea of the problems caused by the direct adulteration of foods with metallic salts a hundred years ago. There were, in addition, numerous references in *The Analyst* to the sort of unintentional contamination frequently met with even today due to inappropriate materials for processing and packaging of foods: lead from tin plate, solder and pewter; tin from unlacquered cans; lead from rubber rings used to close jars. These incidents all occurred well after regulations and laws had been introduced in Britain and elsewhere to prevent adulteration of food sold to the public. How much worse the situation was in pre-Food Act days can be gathered from Accum's *Treatise on Adulterations of Food*.

THE ORIGINS OF BRITISH FOOD LAWS

Britain, like many other countries, had some laws for the protection of the public against unscrupulous food sellers even before the nineteenth century. For many hundreds of years Justices of the Peace had been required, when holding assizes, to inquire into complaints of citizens concerning the quality of bread, beer and other basic foodstuffs sold by merchants. They were to punish those who sold underweight or adulterated products. Probably one of the first food laws in England was the Act to Regulate the Sale of Bread passed in the time of Henry the

Third in 1266. Between then and the nineteenth century many other acts and statutes of a similar nature dealing with specific foods had been passed. However, with the increase in population that followed the Industrial Revolution and the massive shift from rural to urban living, and the consequent increase in demand for the services of middlemen to supply food, the earlier, simpler food laws no longer proved effective in protecting the public from fraud. The new powers of the printed word and pressure generated by what might today be described as the scientific lobby, helped to bring about a much needed reform in food-related legislation. Of considerable importance in making this reform effective were developments that took place in analytical chemistry in the late eighteenth and early nineteenth centuries.

One of the first and most significant moves in the fight against food fraud was the setting up by Parliament of the Select Committee on Adulteration of Food in 1855. The report issued by the Committee led to the passing of a landmark law, the Adulteration of Food and Drink Act of 1860, unfortunately, an ineffective piece of legislation. It was virtually powerless since it left responsibility for maintenance of food quality a local and purely optional one. The law was improved in 1872 when a second Act of the same name was passed. This was of considerable significance since it made the appointment of public analysts, with duties to test foods, mandatory. However, the law was still relatively ineffective and lacked real powers of enforcement. It was only in 1875, after another select committee had been appointed, that the powerful Sale of Food and Drugs Act was passed. As has been mentioned already, this Act is the basis of all modern British food laws. The Act had teeth. Offenders could now be fined for the adulteration of food. Almost immediately, the quality of food sold to the public began to improve and a decrease in the amount of food adulteration was noted. After an initial burst of activity, a rapid and sustained decline in the number of prosecutions brought under the Act followed.

The 1875 Act was amended and replaced by a series of other Acts during the following half century, until 1955 when the present Food and Drugs Act was passed. All foods sold in Britain today are subject to the general regulations of this Act. Under it, government ministers have powers to make regulations covering the composition of foods, including additives and contaminants. Two advisory committees assist the ministers in preparing regulations. These are the Food Standards Committee (concerned with the composition, labelling and advertising of food) and the Food Additives and Contaminants Committee (which advises on the need for, and the safety in use of, additives as well as levels of contaminants permitted in foods). Regulations are made after discussion with interested bodies and individuals, including scientists, industrialists and the consuming public. The laws when promulgated are enforced by food and drug agencies of the different local authorities.

THE ORIGINS OF FOOD LAWS IN THE USA

Though food laws in the USA may operate in a somewhat different manner from those of the UK, both sets of laws have to some extent a common origin. As Alexander M. Schmidt, Commissioner of Foods and Drugs, US Department of Health, Education and Welfare, remarked at a symposium held to celebrate the hundredth anniversary of the UK Sale of Food and Drugs Act, there is a distinct resemblance between the food laws of the two nations[9].

In the colonies of New England the law was that of England, with 'assizes of bread', for example, in keeping with the practices of the mother country. However, with the development of a spirit of independence, local laws dealing with specific New World problems began to be enacted. Since many of these laws came in response to the needs of producers and consumers in the 13 separate colonies and, later, states, a diversity of legislation emerged. One law, passed in 1723 in the Massachusetts Bay colony, prohibited the distillation of rum in stills made from lead. This prohibition rose directly from the realisation by the colonialists that the painful affliction of rum drinkers which was known as 'dry gripes', similar to the 'Devonshire colic' of cider drinkers in England, was caused by lead contamination of the beverage. It is interesting that some 200 years later 'dry gripes' was again to be experienced by American drinkers who consumed lead-contaminated illicit 'bath tub' gin in Prohibition days.

A post-Independence law passed in Massachusetts in 1784, the first general food law in the USA, showed that New Englanders took the question of food quality seriously, though it is doubtful if this law was any more effective than earlier legislation in stopping abuses. The new law provided for the prosecution of 'evilly disposed persons, who, from motives of avarice and filthy lucre, have been induced to sell diseased, corrupted, contagious or unwholesome provisions to the great nuisance of public health and peace'.

The separate states continued to pass their own food laws during succeeding centuries and attempts to promulgate national laws in the federal Congress were not widely supported. As was noted also by Schmidt at the 1976 London Symposium, 'many Americans, including members of Congress, thought that food and drug protection, if needed at all, was a function for the States, and most States already had laws dealing with such matters'. Nevertheless, vigorous attempts were made by many to have comprehensive food laws enacted by Congress, especially after the passing of the UK laws of 1860 and 1875. More than 100 bills were introduced into Congress in the last two decades of the nineteenth century, but with little success. Only a few minor laws were passed, including one requiring inspection of tea and another setting standards for butter and margarine. It was not until 1906 that a comprehensive federal food law was enacted.

The 1906 law defined the legal meaning of 'foods' and drugs' and prohibited their introduction into inter-state commerce if misbranded or adulterated. It also made illegal the use of certain harmful chemicals in food. The law marked a great step forward, but several weaknesses in the legislation appeared in subsequent years.

Another milestone in the development of US food laws was the Shirley Amendment of 1912 which prohibited the labelling of medicines with false therapeutic claims. In the 1930s public concern grew once more about widespread sales of adulterated foods and spurious medicinal cures and drugs. Upton Sinclair's novel The Jungle, did much to stir up this concern. Public pressure for government action continued and in 1938 the well designed and effective Federal Food, Drug and Cosmetic Act was passed by Congress. This is still the foundation of modern American food law. The Act provided, among other things, that the Food and Drug Administration could set standards for the composition of foods. It also provided for ingredient labelling of foods not standardised. It prohibited the addition of poisonous substances to foods and set limits on permissible levels of toxic substances which could not be totally excluded from certain foods. Some of the provisions of the Act were adopted from previously enacted regulations which had been in force for many years in some of the states which were major food producers.

The 1938 law was not long in operation before amendment was needed to cover unforeseen loopholes and allow for new developments in food technology and science. In addition it began to be accepted by many that the absence of a requirement that the safety of food additives should be demonstrated before their use was permitted, was a weakness in the legislation. Under the law any chemical that was in use could continue to be used until the government proved in court that it was harmful, and the burden of proof was entirely on the government. In 1949 a select committee of the House of Representatives, the Delaney Committee, was appointed to investigate the use of chemicals in foods and cosmetics. Out of its deliberations came several amendments to the Act. The Committee adopted the principle that the safety of ingredients should be determined before the public is exposed to them. This was a fundamental change in legislation. It made the manufacturer responsible for establishing the safety of his products, with the government responsible for evaluating the evidence submitted.

Mandatory pre-marketing clearance, as required by the above amendment, has been the subject of much criticism. Even more criticism has been levelled at a further amendment known usually as the Delaney Clause (Food Additives Amendment of 1958, FL 85–929, and Colour Additive Amendments of 1960, FL 86–618). The Clause banned from use, as an additive in commercially formulated foods, any substance which was found 'to induce cancer in man or animal'. The Clause has been called, by its opponents, unscientific in that it banned substances for use as food additives without considering their 'no effect' levels, and wasteful in that it forbade balancing benefits of use against risks. Its defenders have been vocal in their support of the Clause, pointing out that it did not deal with trace amounts of substances that might be consumed by individuals, but only with batch amounts used as intentional additives commercially. In fact, the Delaney Clause was seldom invoked to ban an additive. The Food and Drug Administration adopted the policy of seeking to curb the use of suspected carcinogens through application of the general requirements and not specific amendments of the Act[10]. Its indirect influence, however, has been considerable. In principle, the Clause placed a growing number of suspected carcinogens in the same class of cumulative, chronic health hazards as radionuclides and subjected their use to similarly stringent restrictions[11].

INTERNATIONAL STANDARDISATION AND HARMONISATION OF FOOD LAWS

The history of the development of food laws in many other technologically advanced nations is very similar to that of the UK and the USA. The change from home production of basic foodstuffs to dependence on shops and markets which followed industrial development and urban expansion, led to the passing of many specific and, later, comprehensive food laws in Germany and other countries of Continental Europe. The major principles underlying all such legislation was the protection of health of the consumer and prevention of fraud. Though sharing a common foundation, national differences in outlook and diversity of economic interest led to considerable variation between the actual laws passed in the different nations. As a result there was, and even today within the EEC continues to be, a great variety of laws and regulations governing the composition, labelling, handling and other aspects of food production and sale. These differences can act as a barrier to free trade between nations. As the world grows smaller in terms of travel, communication and international trade, the need for reduction of such barriers becomes even more evident. Pressure has built up for harmonisation of national legislation relating to foods and for the development of internationally acceptable food standards. The United Nations, through two of its agencies in particular, the Food and Agricultural Organisation (FAO) and the World Health Organisation (WHO), as well as on regional bases, the European Economic Community and other international groupings in the Americas and the Pacific, have been actively taking steps in recent years to meet this demand.

EEC REGULATIONS

When the European Economic Community was established under the Treaty of Rome in 1957, one of its principal aims was the facilitation of trade between Member States. In spite of the fundamental relationship between the food laws in the different nations of Europe, the intricacies of legislation in each of them made freedom of movement of goods and expansion of inter-State trade in food difficult. Consequently, the Community instituted a programme of harmonisation which involved the formulation of legislation that would be generally applicable within the EEC.

As might be expected, the machinery for dealing with food legislation within the EEC is complex, far more so than that involved in the development of laws in any one nation. Different bodies are involved: the Assembly or European Parliament, which plays a mainly consultative role; the Council of Ministers, consisting of one minister from each Member State, which acts as the major decision-making body; finally, the chief administrative and executive organ of the EEC, the Commission. Members of the Commission of the European Communities, to give it its official title, are appointed by mutual agreement of governments of the Member States. The Commission draws up rules and regulations for submission to the Council. Both Commission and Council are assisted by a number of advisory and consultative committees.

Under Article 100 of the Treaty of Rome, the Council gave the Commission responsibility for initiating an action programme to harmonise food legislation. This was to be achieved through Regulations, which are laws binding in every respect and replacing equivalent domestic legislation, and Directives, which also bind Member States as to the final result to be achieved, but leave them free to choose the method to be used in attaining this end.

Such EEC legislation must pass through a number of distinct stages before it comes into force. Initially, a draft is drawn up by the Commission in consultation with government experts from the Member States and representatives of industry and consumers. The proposed legislation is then submitted to the Council and is also considered by the European Parliament as well as by the Economic and Social Committee. Finally, if it has been found acceptable to all parties, the legislation is promulgated by the Council. Even then, Regulations and Directives of the EEC do not immediately become law in Member States. In the UK, for instance, Directives can only be implemented when a Regulation is made under the Food and Drugs Act.

Development of harmonised food legislation in the EEC is, clearly, a slow and painstaking process. As a consequence, only a relatively small amount of legislation has been passed in its final form up to the present time. Much remains in draft form. However, though time-consuming, it is a worthwhile exercise and its eventual benefits to the food industry, in Europe as well as worldwide, will be considerable.

A good example of how the EEC operates in an area of food legislation is given by Rossi in a recent review[12]. The particular piece of legislation arose out of a 1973 Council Resolution to initiate an action programme for harmonisation of legislation concerning materials and articles intended to come into contact with foods. It is a particularly difficult area for harmonisation since almost all the Community countries have their own, often differing, regulations on monomers, plasticisers, colouring agents, heavy metals, and other substances suspected by the community of being harmful. Moreover, the general approach to legislation in the area differs between countries. While most favour the use of positive lists or lists of permitted substances, together with various restrictions on their use in plastics, the UK has instead opted for a system which combines a basic statutory framework supported by a monitoring programme and encouragement of industry to take voluntarily any necessary remedial action[13].

A long series of Directives, dating from 1976 when the 'framework Directive' was adopted, to the present, mark the slow process of development of the harmonised legislation. Of particular interest in the context of this book is Directive 82/711 which, among other items, deals with specific migration limits for lead and cadmium, as well as the procedures for performing tests for such migration. As Rossi notes, 'an objective review of the results obtained in 14 years of activity can lead only to the conclusion that progress with the action programme has been very slow for various reasons. The most important of these are the technical complexity of the problems involved and the Community's decision-making procedures which, because they call for unanimity, tend to generate divergence, rather than convergence of opinion'.

It is to be hoped that concerted effort and good will on the part of Member States will result in an acceleration of harmonisation in anticipation of the 1992 removal of internal barriers to trade. A hopeful comment, made from the UK point of view, is that of McGuinness in the paper referred to above: 'assuming the harmonization on food contact plastics remains an EEC priority and that there is good will on all sides, there will be detailed Community regulations in this area implemented into UK law within the next few years. The UK government continues to work towards this objective, attempting to be as flexible as possible, but success will depend on Member States showing a similar commitment to progress. The UK recognises that it will have to change its whole approach to legislation for there to be harmonization. Similarly, other Member States will need to make fundamental changes of their various national legislations. The goal of truly harmonized legislation is an important one for the whole Community although it is one that will not be reached easily'.

THE CODEX ALIMENTARIUS COMMISSION

The establishment of economic groupings such as the EEC and similar organisations in Africa, the Americas and Asia, which are aimed primarily at creating common markets among Member States, has stimulated interest not only within the groups, but also outside them in the need to remove economic and other barriers to international trade in foodstuffs. Practical interest in such moves has been shown by member governments of the FAO and WHO. Initially in 1958 an FAO committee of experts collaborated with the International Dairy Federation to produce a Code of Principles on Milk and Milk Products. In the same year a body known as the Codex Alimentarius Europaeus was established by the International Commission on Agricultural Industries working in collaboration with the Permanent Bureau of Analytical Chemistry. The aim of the new Codex body was the elaboration of internationally acceptable food standards. In 1961 the work of the European Codex was, by mutual agreement, taken over by the FAO and WHO. A joint FAO/WHO Food Standards Programme was established with the Codex Alimentarius Commission as a subsidiary body. The Commission has had considerable success in achieving its objective and the standards it has developed are being adopted in national food legislation by many countries.

The objectives of the Commission are to develop international food standards and to publish these as the *Codex Alimentarius*. In the three decades since the Codex was first established a great deal of work has been done on compositional, labelling, additive, contaminant, residue, hygienic, sampling and analytical aspects of food. In collaboration with member governments, international agencies and other bodies, and following a detailed and prolonged procedure consisting of eight separate steps which allow adequate consultation with interested parties, a number of standards for various foodstuffs have been proposed by the Commission. Proposals are based on recommendations of specialist committees, assisted as necessary by expert working groups. The standards are presented in a uniform manner and are intended for international adoption. The basic aims of all the standards are in keeping with most food laws, to protect the health of consumers and ensure fair practices in the food trade.

The Commission, through its standards and other publications, plays an important role in promoting international harmonisation of food laws. It invites governments to accept its standards, either fully or with minor variations, or only after a target date is reached. Considerable support is being given by many countries to the work of the Codex Alimentarius Commission. Its views and proposals are taken into account in the preparation of food legislation in the EEC, as well as in individual countries such as the UK and Australia. National food standard committees, and other related expert committees, draw on the Commission for information and advice. As we shall see, Australia is one such country which has sought to bring its food laws into line with Codex standards.

FOOD LEGISLATION IN INDIVIDUAL COUNTRIES

In the first edition of this book, an attempt was made to summarise legislation relating to metals in food then current in a number of different English-speaking countries. It was noted, however, that the summary was of little practical use and that 'consultation of the original laws and statutory regulations cannot be omitted by anyone who wants the full picture'. The hope was expressed that, eventually, the multiplicity of individual laws would be replaced by 'a universally applicable and acceptable Codex covering food standards and legislation'.

Ten years later we are still waiting for international acceptance of such a Codex though, in the past decade, appreciable progress has been made in the direction of international harmonisation of food laws. One of the most significant advances has been the decision of several governments to revise their national food laws in conformity with the standards and guidelines of the Codex Alimentarius Commission. Australia has been one of the first to do so and her recent Model Food Act could serve as an example for other nations and not just for the individual states in the Australian Federation. For this reason, Australian Food Standards will be explained in some detail, with particular reference to those that apply to metals in food. No attempt will be made to summarise comparable legislation at present in force in other countries.

FOOD STANDARDS IN AUSTRALIA

Australia, a country approximately the size of the United States of America, but with a population of some 17 000 000 people, is a federation of six separate states and two territories. Each state and territory has its own parliament and passes its own laws relating to internal affairs. There is also a federal parliament at Canberra, the national capital, responsible for external affairs and defence as well as for a multiplicity of national and inter-state matters.

Public Health/Food Legislation is the responsibility of the individual state and territory parliaments. Until recently most of the provisions in such laws relating to food had been passed in the first decade of the century. Though many of the laws and regulations had been modelled on the UK 1875 Sale of Food and Drugs Act and its subsequent Regulations, local interests and inter-state rivalry resulted in considerable diversity of legislation between states. Differing requirements and standards proved to be an impediment to free trade in foodstuffs between states and resulted in increased costs to manufacturers and consumers. It was widely recognised that this was an unsatisfactory situation and for many years suggestions had been made at both state and federal level that steps should be taken to rationalise food laws on a national scale. It was not, however, until 1975 that it was agreed at a conference attended by health ministers from each state and territory, as well as from the Commonwealth Government, that a Model Food Act, suitable for adoption by states and territories should be developed.

Several years of planning and consultation, coordinated by the Commonwealth Government and with input from all states and territories, as well as from non-government bodies and individuals, followed. Food laws of each of the states and territories were considered, as well as food legislation of the UK, Canada, New Zealand, Ireland and other countries. Particular attention was paid to the FAO/WHO Model Food Act.

By 1980 planning and development were completed and the text of the proposed Model Food Act was presented to the Health Ministers Conference which endorsed the Act. The Ministers agreed unanimously to initiate procedures for its adoption in their respective jurisdictions. Adoption proceeded, with some difficulties and delays, until the present time when the Model became the law in most states and territories.

The Food Act, in fact, contains very little in its text about food. What it does is to empower authorised officers or departments to take certain actions related to the purpose of the Act ('securing the wholesomeness and purity of food', as the New South Wales Pure Food Act states[14]) such as the making of regulations prescribing standards for food.

The launch of the Model Food Act was followed by drafting of Model Food Standard Regulations, also for adoption by individual states and territories. The responsibility for preparing these regulations has been given to the National Health and Medical Research Council, which brings together in its committees the expertise of the Commonwealth, the states and territories, industry, consumers and independent experts in matters relating to food and health. As well as preparing the initial Model Regulations, the NH&MRC also has an on-going role of monitoring developments and recommending changes to food standards when necessary.

TABLE 10
Maximum Permitted Levels of Metals in Foods (Queensland)

Metal	Food	Maximum level (mg/kg)
Antimony	Beverages	0·15
	Other foods	1·5
Arsenic	Beverages	0·1
	Fish (inorganic As)	1·0
	Other foods	1·0
Cadmium	Beverages	0·05
	Fish	0·2
	Liver	1·25
	Offal (not liver)	2·5
	Other food	0·05
Copper	Beverages	5·0
	Offal (not ovine liver)	100·0
	Ovine liver	200·0
	Other foods	10·0
Lead	Beverages	0·2
	Meat/fish (in tinplate)	2·5
	Vegetables	2·0
	Other food	1·5
Mercury	Fish, molluscs	0·5 (mean)
	Other food	0·03
Selenium	Beverages	0·2
	Offal	2·0
	Other food	1·0
Tin	Canned-selected fruits, and vegetables	250·0
	Other food-in tinplate/tinfoil	150·0
	Other food	50·0
Zinc	Beverages	5·0
	Oysters	1 000·0
	Other food	150·0

Several states have now introduced legislation for the automatic adoption of Commonwealth Food Standards under their individual Food Acts. This is extremely important if a return to the former lack of harmony between food laws and regulations across Australia is to be prevented.

Metals in Food Regulations (Queensland)
The State of Queensland has, like most other States of Australia, adopted the NH&MRC Food Standards Code under its Food Standards Regulations of 1987, with the words: 'every provision of the Code shall be

deemed to be a provision of these regulations and to be prescribed by these regulations and shall have force and effect accordingly'[15]. A section of the Regulations of particular interest is that on *Metals and Contaminants in Food*.

The section begins with the note that the word 'metal' includes compounds of the metal; that antimony, arsenic and selenium are deemed to be metals, and that maximum permitted concentrations shall be determined on the edible content of the food.

The Regulation is presented in the form of a table of three columns, the first listing the metal, the second specifying a particular type of food, and the third the maximum permitted concentration of the metal in the food listed. It is summarised in Table 10, with several of the food types listed in the original document omitted.

CONCLUSION

International harmonisation of food laws is continuing to take place. The pace is slow but, undoubtedly, commercial pressure will eventually be successful in bringing it to completion. As more and more countries adopt the FAO/WHO *Codex Alimentarius* Model Food Act and Food Standards, as Australia has done, differences between food laws will largely disappear across national boundaries. It will then no longer be necessary for food manufacturers and traders to check whether permitted levels of a metal or other adventitious components of his product are the same in his own country and in that to which he plans to export food. However, for the present, he cannot assume that such uniformity exists, and it is still necessary for him to consult the appropriate regulations. The most recent edition of Lenane's *Summary of Food Standards and Regulations*[16], or some similar reference volume must still be consulted.

REFERENCES

1. COMMONWEALTH DEPARTMENT OF HEALTH AUSTRALIA (1986). *Model Food Legislation* (Australian Government Publishing Service, Canberra).
2. ACCUM, F. (1820). *Treatise on Adulterations of Food and Culinary Poisons and Methods of Detecting Them* (Longmans, London: reissued by Malkickodt, USA, 1966).
3. CROSBY, N. T. (1977). Determination of metals in food. *Analyst*, 102, 225–68.
4. MAYRHOFER, J. (1891). Copper in preserved foods. *Chem. Zeit.*, 15, 1054.
5. BODMER, R. and MOOR, C.G. (1987). On copper in peas. *Analyst*, 22, 141–3.

6. ALLEN, A. H. and COX, F.H. (1897). Note on the presence of heavy metals in cheese. *Analyst*, 22, 187–9.

7. WOLFF, J. (1898). Rapid method of detecting lead chromate in papers used for wrapping eatables. *Ann. de Chimie Analyt.*, 2, 105.

8. KAISER, H. (1899). Presence of barium salts in Cayenne pepper. *Chem. Zeit.*, 23, 496.

9. SCHMIDT, A.M. (1976). The development of food legislation in the U.K., in *Food Quality and Safety: A Century of Progress. Proceedings of the Symposium Celebrating the Centenary of the Sale of Food and Drugs Act 1875*, 4, London, October 1975 (MAFF/HMSO, London).

10. HUTT, P.B. (1978). *Food Drug Cosmet. Law J.*, 33, 501–92.

11. TURNER, J.S. (1970). *The Chemical Feast* (Grossman, New York).

12. ROSSI, L. (1987). *Food Add. Contam.*, 5, 21–31.

13. MCGUINNESS, J.D. (1986). The control of food contact plastics — the next ten years. *Food Add. Contam.*, 3, 103–12.

14. FOOD INFORMATION SERVICE (1989). Food legislation. *Food Australia*, 41, 903–4.

15. QUEENSLAND GOVERNMENT. *Food Standards (Adoption of Food Standards Code and General) Regulations* 1987. *Queensland Government Gazette*, 29 August 1987.

16. LENANE, G.A. Summary of food standards and regulations, in *Food Processing and Packaging Directory* (Consumer Industries Press, London).

CHAPTER 4

METAL ANALYSIS OF FOOD

The past decade has seen many improvements in techniques and equipment for the analysis of metals in foods, especially at trace levels of concentration. It is not that a wide range of totally new instrumental techniques has become available. Rather, methods and equipment which were already known to the analyst, and in use in analytical laboratories, were refined and improved, allowing an increased sensitivity, reliability and speed and the ability to analyse even complex materials with a higher level of accuracy than had previously been possible. As a result it is now well within the means of even a relatively inexperienced analyst, with the facilities available in a moderately equipped laboratory, to tackle trace elements such as chromium and selenium which were, until recently, considered to be the preserve of the specialist.

As will be noted below, two other factors, apart from improved instrumentation, have been of considerable significance in facilitating these advances: the availability of an increasing range of certified reference materials for monitoring precision and accuracy of methods, and wider acceptance of the basic truth that trace metal analyses of foods and other biological materials required careful planning, meticulous attention to cleanliness and use of validated and reliable techniques.

How restricted food analysts were, with regard to determination of metals in foods as recently as the late 1970s, was noted by Stewart in a review of modern methods of analysis[1]. Stewart classified methodologies for determining nutrients in food according to the probability of obtaining a correct value, the speed of analysis, and cost, using five descriptive terms: (1) sufficient, (2) substantial, (3) conflicting, (4) fragmentary, (5) little to none. His ratings for the methodology of copper and zinc were (1); for selenium (2); for arsenic, chromium and manganese (3); for cobalt, iron, silicon, tin and vanadium (5).

About the same time an inter-laboratory study[2] found that results for chromium in brewers' yeast, reported by 14 different participants, ranged from 0.351 to 5.40 μg/g for identical samples. Another inter-laboratory study[3], of trace elements in milk powder, found the following range of results: As 5–544 ng/g; Cd 1–1660 ng/g; Co 0.004–51 μg/g; Cr 0.02–52 μg/g; Fe 0.70–20 μg/g; Hg 1–666 ng/g; Mn 0.12–55 μg/g; Mo 0.15–3.6 μg/g; Pb 0.04–246 μg/g; Se 25–313 ng/g; V 0.002–20 μg/g; and Zn 26–4149 μg/g. Such results, as well as the wide range of findings reported in the literature for various metals in biological samples and food, did little to reassure the less than dedicated analyst that reliable results could be obtained with the equipment and techniques available in the ordinary laboratory. It was even more alarming to read that the Advisory Group to the 1978 Symposium on Nuclear Activation Techniques in the Life Sciences, estimated that more than half of the approximately 16 million trace element determinations being performed each year were inaccurate and unreliable[4].

The situation has changed for the better, though problems still occur. Accurate and reproducible analytical results for many of the metals can be obtained by a careful scientist who has access to modern equipment, not all of which is unreasonably expensive. Several recent textbooks and manuals are available to help in this task. Among the most enduring and for many situations that of choice, is the Official Methods of Analysis produced by the Association of Official Analytical Chemists, Washington, DC. The fourteenth edition was published in 1984. Among textbooks of particular value are those by Stewart and Whitaker[5], the CRC *Reviews in Analytical Chemistry*[6], and such specialist books as Marczenko on spectrophotometric methods of analysis[7] and the CRC *Handbook of Atomic Absorption Spectrophotometry*[8]. The text by Versieck and Cornelis[9], though its title indicates concentration on plasma and serum, is an excellent, up to date and instructive study, far broader than might appear at first glance.

It is not intended to offer here a substitute for these and other works, but rather to gather into one section a brief treatment of a number of methods of analysis of metals in food, to draw attention where necessary to particular problems and to provide a readily consultable outline of the topic. This is not primarily intended for the professional analyst or the skilled laboratory expert, but for that breed of scientist who has little more than an undergraduate foundation in analytical methods and needs in the course of his investigations to analyse foods for their metal components. Many of the procedures described have been in use in the author's laboratory and have been found to be well within the competence of the less experienced analyst. For these, as well as for the more experienced food scientist, it is hoped that this section, though limited in scope, will be welcome.

GENERAL METHODS OF ANALYSIS: SAMPLING

Proper sample collection, as well as storage and preparation of material for analysis, is essential if errors are to be avoided. The need for care in pre-analytical procedures has been stressed again and again by experts in the field[10]. It has been noted[11] that the fall in reported values over recent years, for example in plasma trace elements, 'is primarily due to more careful sampling procedures'.

Obtaining a Representative Sample
Foodstuffs are seldom homogeneous in nature. Even liquids can be stratified in layers of different concentrations. It is essential that small samples taken from a larger bulk sample should be representative, if analytical results obtained on the small quantities are to be applied to the whole. Sampling must be carefully carried out to ensure that errors due to the inherent inhomogeneity of the bulk foodstuff are minimised. The same must be done even with less bulky samples when, as is often the case with animal and plant tissues, the nature of the material is naturally heterogeneous. In some cases the problem may be solved by homogenising the total sample, but this may not be a suitable solution in the case, for instance, where only the edible portion of a plant or animal foodstuff is of interest. The problem caused by sample matrix variation, and methods for its reduction by unit compositing, has been discussed by Lento[12]. Bulk material is normally selected for analysis either by random or by representative sampling. When random sampling methods are used, portions are taken in such a way as to ensure that every part of the material has an equal chance of appearing in the sample. It is important to avoid bias due to personal preferences of any sort. This can be done by numbering the containers and then using a table of random numbers to control selection by numbers, irrespective of external appearance or any other extraneous factor. A widely used source of random numbers is Table XXXIII in the classic *Statistical Tables* of Fisher and Yates[13] which was first published in the 1930s but has been reprinted in many modern books on statistics. It may also be necessary to carry out subsampling subsequently by quartering or other methods if the containers selected by the use of random tables are too big.

Large bulk containers may have to be subjected to representative or stratified sampling. In this procedure, samples are taken in a systematic way so that each portion selected represents a corresponding portion of the bulk. The container is considered to be divided into different sections or strata and samples are removed from each stratum. The size of the samples removed should correspond to the relative proportions of the imaginary strata of the bulk. Thorough mixing, for example by end-over-end rotation, is preferable to representative sampling in the case of liquids and sometimes even of very fine powders.

CARE IN SAMPLE PREPARATION AND HANDLING PROCEDURES

Prevention of Contamination of Samples

It is of the greatest importance that sampling and handling be performed so as to avoid all possible contamination or loss. It is not unknown for less able analysts to concentrate all their care and efforts on optimising their instrumental techniques and to neglect pre-analytical factors which are crucial to the success of their efforts[14]. Kumpulainen and Paaki[15] have drawn attention to this problem in a paper on the quality control programme adopted by the sub-network of the FAO–European Co-operative Network on trace elements in foods and diets. Among the procedures they list are the following:

—Homogenisers used in sample preparation must have blades made of titanium (99.4 per cent pure) or other material which will not cause trace element contamination.

—Knives and corers used should also be made of similar quality titanium.

—Another procedure which can reduce the possibility of metal contamination during this stage of preparation is the use of ceramic cylinders to pulverise the sample[16]. Sample containers should be made of either polyethylene or polypropylene and should preferably not have been used previously.

—Final preparation of samples and pooling should be carried out by operators wearing talc-free polyethylene gloves, under a clean-air hood.

Schmitt's paper[10] similarly stresses the need for particular care to prevent contamination during sample preparation. It notes among the typical sources of error several factors relating to the environment of the analytical equipment, including laboratory staff skin, perspiration, cosmetics, cigarettes, clothing and the local atmosphere. Dust in laboratory air has been found to contain 3 g Al, 1.6 g Zn and 0.2 g Cu per kg[17]. Even clean-room conditions may not always be adequate, as Cornelis and Schutyser[18] noted. They found 8 $\mu g/m^2$ of Al in a normal laboratory and 0.9 $\mu g/m^2$ under clean room conditions. Not many laboratories today would tolerate cigarette smoking by staff while engaged in analytical work, but it still known to take place on occasion. Cigarette smoke has been found to contain 7.3 $\mu g/g$ Fe, 90 mg/g Si and 10 $\mu g/g$ Zn[19].

Drying of Samples

Where analytical results are required in terms of fresh food or 'portions as served', samples will often not need to be subjected to any treatment other than homogenisation or grinding. However, if samples have to be stored for any length of time or if 'dry weight' results are required, further preparation is necessary. This can involve a simple drying to constant weight in a fan-assisted air oven, at 70–100° C, care being taken to avoid overheating or charring. A vacuum oven is useful for allowing use of safer, lower drying temperatures and speeding up the process. The European Cooperative Network[14] recommends that freeze drying be used whenever possible for food samples destined for trace element analysis. This is highly desirable, if facilities are available. It reduces the moisture content of the sample to a suitable level and at the same time produces a crumbly texture which is convenient for subsequent subsampling and analysis. The freeze-dried samples, sealed in a plastic, preferably polyethylene, bag can be stored without refrigeration.

Purity of Reagents and Water

Chemical reagents used in procedures described below for sample digestion and solution can be a source of contamination. Even analytical grade reagents can be at fault in this way at times. It is important that those that are used be tested for impurities. Steps may have to be taken to remove contaminants if they are found to be present[20]. Veillon's advice to 'check everything'[21] is most appropriate with regard to reagents used in trace metal analyses.

Water used for sample preparation and other purposes must also be tested for absence of contaminating metals. Only distilled or deionised water purified to a very low conductivity should be used.

Glassware, Other Equipment
All glassware should be thoroughly cleaned to remove all possible contamination. A suitable procedure is to soak glassware overnight in an alkaline detergent (such as 3 per cent vol/vol Decon), rinse with deionised water and follow with a further soaking in 2 per cent vol/vol hydrochloric acid. The items are then washed in deionised water, followed by two rinsings in deionised water and left overnight for drying in a clean room.

PREPARATION OF SAMPLES FOR ANALYSIS

Most of the generally available analytical techniques for the determination of metals in foodstuffs require that samples be prepared destructively before instrumental analysis takes place. Exceptions to this are certain beverages, including water, to which it is possible to apply colorimetric and other analytical techniques directly without pretreatment. Organic matter and other components present in the majority of foods would interfere with the analytical process unless they were removed.

Organic matter is usually removed from a foodstuff by some form of oxidation, either by the use of oxidising acids in a wet digestion or by dry ashing in the presence of air or pure oxygen. The choice of method depends on the metals to be determined as well as on the nature of the food to be analysed. What is aimed at is a procedure which gives reliability and accuracy in the range of concentrations appropriate to the investigation. At the same time the method should permit work to be carried out with reasonable speed, cost effectively and with the facilities available to the investigator. The method actually chosen is often a compromise, for, of the many techniques and procedures reported in the literature for the different metals, it is often found that no single method will have all the desirable qualities at the same time.

Sample Digestion
In the following section a variety of procedures for the digestion of food samples prior to instrumental analysis will be described briefly. Many of them have been found to be effective in the author's laboratory; others are reported in the literature for which references are given.

Wet digestion and dry ashing are currently the most common procedures used for the decomposition of foods which are being analysed for their metal components. In wet digestion strong oxidising acids, such as nitric, sulphuric and perchloric acids, are used, either alone or in various combinations. In dry ashing samples are incinerated at a suitable temperature in a muffle furnace and the resulting ash, free of organic compounds, is dissolved in an acid solution.

Sansoni and Panday[22] provide an extensive review of ashing techniques of biological materials. They cover, in addition to the more widely used methods, such techniques as oxidative fusion, vapour oxidation, deductive decomposition, and the use of halogens. The review article should be consulted if details of these techniques are required.

Dry Ashing

This is a convenient method for oxidising organic matter in preparation for instrumental analysis of many, though not all, metals. Loss of several elements, particularly mercury, arsenic and selenium, occurs by volatilisation during ashing. The advantages of the method include the possibility of using larger samples than in wet digestion, the absence in many instances of danger of contamination from added reagents, and the fact that the procedure does not generally require constant attention by the operator. However, dry ashing is a lengthy and time-consuming technique. It also requires careful temperature control if volatilisation losses are to be avoided. A useful appraisal of the technique is given in an excellent and comprehensive, though now somewhat dated, paper by Gorsuch[23]. A variety of approved methods have been published by the Analytical Methods Committee (AMC)[24].

Dry ashing is usually carried out at temperatures between 400 and 600° C but temperature and length of combustion will depend on the particular metal being analysed and the need to avoid loss by volatilisation (especially of copper, selenium, antimony, cadmium, arsenic, mercury and some other elements) or by combination with the materials of the crucible. Excessive heating may also make certain metallic compounds, such as those of tin, insoluble. Some flour products can produce a dark melt in which carbon particles are trapped and will not burn, thus preventing complete destruction of the organic matter.

The use of ashing aids is recommended in certain situations to assist decomposition of organic matter and recovery of metals. A variety of ashing aids are described in the AMC recommendations. They include salts of metals and acids. Ashing aids may be added to the sample at the start of ignition or after decomposition has been partially completed.

The technique has been used effectively for decomposition of large sample volumes of single foods and blended diets[25]. Maher[26], however, found the technique unsuitable for marine biological materials, as foaming and charring occurred, with resultant poor recovery of elements, especially selenium. Roesch[27] found that complete oxidation of certain foodstuffs, and especially of mixed diets, was difficult to achieve, even when ashing aids were used and temperatures were kept below 500° C.

Wet Digestion Techniques

The applicability of acid digestion to a wide range of materials, its rapidity and its generally superior recovery rate, make it a suitable technique for many investigations. However, it has some disadvantages compared to dry ashing. In particular, since only small samples can normally be handled and relatively large volumes of reagents are required, the technique can lead to high blanks and expose samples to contamination. It is, moreover, potentially a hazardous method and requires careful supervision by the operator.

Nitric Acid Digestion

Digestion using nitric acid (70 per cent) has been found to be satisfactory for many, though not for all, biological materials. It has been used successfully for digestion of hair[28]. A technique in which food was reduced to dryness after repeated digestion in the acid, and then re-dissolved in the same acid with added potassium permanganate, was found to be satisfactory for the determination of chromium in food[29]. Lead, chromium and copper have been determined in digests of plant material prepared in a pressure digestion vessel with 3 ml of nitric acid for about 0.3 g of tissue[30]. Good recoveries for cadmium, lead and zinc in biological materials have also been reported for the use of nitric acid alone[31].

There are conflicting reports about the suitability of nitric acid digestion for the preparation of biological materials for selenium analysis. It has been reported to be unsatisfactory for this purpose because of its low boiling point and the risk of drying out at higher temperatures[32]. However, the use of 90 per cent rather than 70 per cent nitric acid improved recovery of selenium as well as mercury in organic matter[33]. Addition of magnesium nitrate to the nitric acid digestion was also found to improve selenium recovery in such materials[34].

Sulphuric–Nitric Acid Digestion

Addition of sulphuric to the nitric acid in the digestion mixture has been found to improve its oxidising power. This technique has been used successfully in the determination of copper, iron, zinc and manganese in food[35]. The presence of sulphuric acid, with its boiling point of 330° C, results in a higher temperature and gives more rapid and complete digestion than does nitric acid alone.

The addition of hydrogen peroxide to the digestion mixture has been found to improve oxidising power for the determination of copper, iron, manganese, zinc and chromium in plant and other biological tissues[36].

There are divided opinions about the suitability of a nitric–sulphuric acid mixture for digestion of biological materials prior to selenium determination. Bunker and her colleagues[37] found that the mixture was adequate for the determination of the element in foods. Others noted that the mixture, even with added peroxide, did not have sufficient oxidising power to bring about total decomposition of certain organic matrices for selenium determination[38].

Use of Perchloric Acid

Oxidising power and speed of digestion is dramatically increased when perchloric acid is added to either nitric acid alone or to a nitric–sulphuric mixture. The resulting mixture is sufficient to dissolve almost all types of organic matter[39]. Perchloric acid should, however, be used with caution, because of the danger of explosion. It should never be used alone[40]. A perchloric–nitric mixture has been found effective for the digestion of plant material, with recoveries comparable to those in dry ashing[41].

There has been some controversy about the possibility that the presence of perchloric acid causes loss of volatile elements, such as chromium and selenium, during digestion[42]. However, Jones[43], who studied various combinations of acid mixtures using perchloric acid, found no significant losses of chromium during the digestion of brewers' yeast.

Addition of Hydrofluoric Acid

The addition of hydrofluoric acid to a sulphuric–nitric–perchloric acid has been reported to improve recovery of chromium in brewers' yeast by removing silicate interference by forming silicon tetrafluoride[44]. A similar procedure has been found useful in releasing chromium from plant material[45].

Chelation, Solvent Extraction and Concentration of Extract

Where it is necessary to remove interference by other substances which may be present in the sample digest or ash, or when preconcentration of an element is required before instrumental analysis, a number of techniques are available.

Chelation, followed by solvent extraction of the metal chelate into an organic phase, may be used. As Ihnat[46] has noted, besides enhancing detectivity, solvent extraction has the added advantage of leading to reduced interferences by virtue of reducing the matrix burden in the final solution. Many different complexing agents are suitable and methods are reported in the literature. A number of ion-exchange resins can also be used for removing interfering metals from solutions in preparation for final analysis.

VALIDATION OF METHODS AND ANALYTICAL DATA

Alvarez, in a very informative chapter in Stewart and Whitaker's text, to which reference has been made a number of times above[47], comments that knowledge of the accuracy of analytical data is the foundation of a laboratory's quality control. But, he adds, accuracy is difficult to evaluate. He noted that unreliable analytical data may be the result of poor methodology, improper instrument calibration, impure reagents, faulty manipulation by the analyst, or a combination of these.

Reference has been made already to the need for care in analytical procedures, especially with regard to control of contamination. Another step that may be taken is to compare data generated on a homogeneous sample by the laboratory's method of choice, with data generated on the same material by another method of known accuracy. For example, selenium in foods could be determined by graphite furnace atomic absorption spectrophotometry (GFAAS) as well as by the spectrofluorometric method.

A second procedure would be to call on the assistance of another reputable laboratory, using the same or other analytical techniques, to analyse subsamples of the same homogeneous material. Results obtained in the food scientist's laboratory for chromium in powdered milk, for example, could be compared with those obtained by neutron activation analysis (NAA) or inductively coupled plasma atomic absorption spectrophotometry (ICP-AAS) by a commercial or research establishment. Since this would not be a regular cost, but only occasional as a check on quality control, the expense involved would be fully justified.

However, there is a possibility that even apparently homogeneous samples may not, in fact, give the same analytical results, even when accurate procedures are used. To avoid the frustration that can occur when data from different methods do not agree, it is essential that a

Standard Reference Material be analysed by the different methods. Every laboratory interested in accuracy of trace metal analyses in foods and other biological materials must have access to, and use such materials as a routine part of quality control.

Reference materials used in quality control of analytical procedures should be similar in matrix to the actual samples routinely analysed in the laboratory. They must also contain the elements of interest in concentration ranges that can be expected in foods and other biological materials of interest.

Several organisations issue a variety of reference materials as primary controls for quality assurance applications. These are available commercially from the International Atomic Energy Agency, Vienna, the National Bureau of Standards, Gaithersburg, Maryland, USA, and other organisations. Among materials of particular interest to food scientists are the following:

> milk powder (IAEA A-11; NBS SRM-1549)
> bovine liver (NBS SRM-1577)
> animal muscle (IAEA H-4)
> plant tissue (NBS SRM-1572: citrus leaves)
> flour (NBS SRM-1567: wheat)
> mixed total diet (IAEA RM H-9).

These materials are issued accompanied by a certificate which indicates the elements for which reference figures are provided, as well as certified values and confidence intervals. Certified reference materials (CRMs) for most, but not all of the metals of interest to food scientists, are available. Both the National Bureau of Standards and the International Atomic Energy Agency, as well as other bodies interested in trace element analyses, continue to investigate suitable reference materials and to produce updated materials, with expanded ranges of analytes, as they become available.

END-DETERMINATION METHODS OF METAL ANALYSIS

The actual method employed for analysis of metals in samples prepared as described above will depend to a large extent on the facilities available to the investigator. In specialised laboratories, where financial support is generous, neutron activation analysis or mass spectrometry may be the technique of choice. For many research workers, and for scientists in commercial and industrial laboratories, analytical techniques will, of necessity, be less costly spectrophotometry, especially atomic absorption spectrophotometry. Indeed, choice of end-determination methods will more often depend on laboratory space available, the number of samples to be analysed, availability of skilled technical assistance, funding and similar restraints than on strictly scientific considerations. Consequently, while we will look briefly in this section at most of the methods available today for analysis of metals in food, irrespective of cost or complexity, our main interest will be in those methods most readily available and not beyond the competence of a scientist with at least moderate laboratory skills and the patience and intelligence to follow instructions from the specialist literature and the pertinent handbooks.

Spectrophotometry
Many different analytical techniques for the determination of metals in foods are available today. For more than a century analytical chemists have been developing methods for analysis of inorganic elements and compounds and many of their techniques have been applied with success to foodstuffs. This required, generally, a good knowledge of and skill in 'wet' chemistry, though instrumental analysis was also developed. Three decades ago spectrophotometric and polarographic methods were normally the most commonly used instrumental techniques for metal analysis in foods. But with the development, first, of flame emission and, most recently, of atomic absorption spectrophotometry, there has been a decline in the popularity of those other methods. Nevertheless, spectro-photometric determination of elements still has its uses and its strong supporters. Some years ago MacLeod[48] listed a considerable number of, then, recent applications of the technique, many of which are still very useful for certain situations. A more up to date text by Marczenko[49] draws attention to the many new sensitive spectrophotometric methods, as well as the new reagents which have been developed in recent years. The author notes that these recent achievements in spectrophotometry, mainly with regard to sensitivity and selectivity, make it still a most useful method in trace metal analysis.

A useful treatment of the subject of spectrophotometry and its predecessor colorimetry, up to the beginning of the atomic absorption era, is given by Sandell[50]. The technique is based on the simple relationship between absorption of visible or near ultraviolet radiation by a solution and the concentration of coloured species in the solution. This fundamental law of colorimetry was first stated by Beer in 1852.

Conversion of a substance into a coloured complex by a reagent is a first step in spectrophotometric determinations. Originally such determinations were carried out in test tubes, using filters to compare visible colours with standard solutions of known concentrations. Instruments using more sophisticated optics were subsequently introduced, to allow the method to be extended beyond the visible range. As Marczenko notes, many spectrophotometric methods are remarkable for their versatility, sensitivity and precision. Even today the basic instrument is still relatively inexpensive, compared to instruments used in other forms of metal analysis. This is a consideration of some significance to many investigators. Moreover, most spectrophotometric techniques are direct and can be used for all the metallic and most other elements, as is made clear in the detailed instructions on such methods in Marczenko's text. Another advantage of spectrophotometric techniques is that they can often be readily automated for routine analysis.

It is expected that, for reasons such as those given here, spectrophotometry will continue to be used, especially in smaller laboratories, for many years to come. Indeed, it is predicted that an upsurge of spectrophotometry may occur due to the use of trace enrichment methods for detection by high performance liquid chromatography (HPLC)[51].

Fluorometry

Metals can be built into certain organic ion-association complexes or fluorescent chelates which have the property of absorbing light of one wavelength and in its place emitting light of another wavelength. When the light is absorbed by the complex ions, they are raised to an excited state and then, after a short delay, return to the ground state. The photons released are equivalent to the difference in energy between the ground and the excited state. This emission, or fluorescence, is in all directions. For practical purposes of analysis, the number of photons emitted at right angles to the incident beam is measured. The intensity of the re-emitted radiation is related to concentration.

Spectrofluorometry is frequently used for the determination of trace elements in foods and other biological materials. The method is very sensitive, several orders of magnitude better than that found with spectrophotometry. However, the concentration range is limited. It has been found especially useful for the determination of selenium at low levels of concentration in biological materials such as blood. The sample is digested in an acid mixture and a fluorescent chelate formed by adding diaminonaphthalene (DAN)[52].

Polarographic Methods

Though this is a technique requiring greater skill than does spectro-photometry, and usually more elaborate facilities, polarography was widely used before the development of atomic absorption spectro-photometry. It is still used to some extent, especially in the related technique of anodic-stripping photometry. Both techniques rely on the fact that different metals require the application of different potentials before they are deposited from solution onto a cathode. A characteristic half-wave potential results for each metal and can be used for identification, while the height of the wave is related to the concentration. The technique is sensitive if used with care. It is especially useful for the simultaneous determination of several elements, particularly heavy metals. Organic matter must be destroyed prior to analysis. Oxygen must be removed from the solution by bubbling with an inert gas, such as nitrogen, during analysis. Temperature must also be controlled to within a half a degree.

Anodic stripping is a sensitive variation of the polarographic technique. It is especially suitable for such metals as zinc, cadmium, copper and lead and others that have suitable electrode potentials. After the metal ions have been preconcentrated from the solution and amalgamated into a stationary hanging mercury drop electrode, the process is reversed by using a potential more negative than the reduction potentials of the metals being determined. The metals are oxidised and stripped anodically, using a slowly increasing positive potential. The current recorded during the stripping is directly related to the concentration of the metal. The technique can be as sensitive and reliable as atomic absorption spectrophotometry.

Atomic Spectroscopy

When a metal or its compound is introduced into a flame, an atomic vapour of the metal is produced and coloured light is seen. Talbot, in the early nineteenth century, was among the first to put this property of elements to a scientific use when he began to study flame spectra. In the 1860s Bunsen and Kirchoff discovered the existence of caesium and rubidium through their flame spectra. These spectra are still used today and form the basis of a very important family of analytical techniques, atomic spectroscopy.

In the exothermic reaction of two gases, such as hydrogen and oxygen, which results in the production of a flame, the initial reaction is in fact endothermic and involves the breaking of bonds in the gas molecules. This reaction produces free radicals and when these come into contact with an element, or a compound, energy will be released. Some of this energy will be retained within the system and cause a chain reaction, with further dissociation of incoming gas molecules. Collisions between the free radicals first formed and atoms of the element also produce free radicals from these atoms. If sufficient energy is available the new free radicals may be raised to an excited state in which the valence electrons of the atoms are raised to a higher than normal energy level. The excited atoms then drop back immediately to their original energy level, with the atoms returning to their 'ground state'. As this occurs radiation characteristic of the element is emitted. This is the physical basis of emission or flame spectrophotometry.

Even in a very hot flame, only a very small number out of all the atoms of the metal will be raised to an excited state. However, the unexcited remainder can be made to absorb radiation of their own specific resonance wavelength (in general the same wavelength at which they would emit radiation if excited) from an external source and in this way reach the higher energy state. In other words, if light of this wavelength is passed through a flame containing atoms of the element, part of the light will be absorbed, and the degree of absorption will be proportional to the density of the atoms in the flame. This phenomenon is the basis of atomic absorption spectrophotometry[53].

Both emission and atomic absorption spectrophotometry use basically the same type of equipment and many instruments have dual functions. Essential requirements are: (1) a means of atomising the test sample — the atomiser (a flame or an electrically heated graphite tube) supplying thermal energy to the sample which is vaporised and, on further heating, dissociated into free atoms; (2) for atomic absorption only, a stable light source emitting the sharp resonance line of the element to be determined. This can be a hollow cathode lamp (HCL) or an electrodeless discharge lamp (EDL), both containing the element to be determined. Single element lamps are the most commonly used, but multi-element lamps are also available for a number of elements. Requirement (3) is the optical and electronic equipment for selecting the particular energy wavelengths of interest and recording levels of emission or absorption of light.

Atomic Absorption Spectrophotometry
Atomic absorption spectrophotometry, introduced by Walsh at the Commonwealth Scientific and Industrial Research Organisation (CSIRO) in Australia in the mid-1950s, has had a more rapid growth than any other

analytical technique. It is still one of the best and most widely used analytical methods for the determination of major and minor inorganic elements in agriculture, biology, medicine, food studies, mining and environmental studies[8]. It is a relatively easy technique to learn by scientists who have not been trained formally in analytical chemistry.

More than 60 metallic elements can be determined, in a wide range of concentrations, by the method which is used in different modes. Flame atomic absorption spectrophotometry (FAAS) is rapid and is sufficiently sensitive to permit the determination of most of the trace elements in food at the mg/kg range[54]. Graphite furnace AAS (GFAAS) allows determination of a wider range of elements and easily extends into the μg/kg range of concentrations. Some elements are better handled by hydride generation AAS (HGAAS), which involves introduction of gaseous compounds of an element such as arsenic or selenium, into a flame or electrothermally-heated graphite tube. Another method, cold vapour AAS (CVAAS) used for the determination of mercury, is based on the generation of elemental mercury vapour at room temperature.

While GFAAS overcomes many of the limitations of restricted sensitivity and sample size of FAAS, it suffers significantly from matrix effects which are less important with the flame mode. To overcome this problem it is necessary to employ background correction with GFAAS. This can be done by use of a continuum source, usually a deuterium lamp in the UV region and tungsten iodide in the visible.

Zeeman Effect Background Correction

In the Zeeman method of background correction a magnetic field, applied to the source or the atomiser, is used to split the resonance line into its Zeeman components (π and $\pm\sigma$). In effect the single beam is converted into a double beam and polarisers are used to monitor the background on the wings, while the analyte signal and background together are monitored with the central component. The Zeeman method is one of the most effective techniques for background correction and has enabled considerable advances to be made in trace metal analyses in recent years[55].

Atomic Emission Spectrophotometry

AES is based on the spectral line radiation emitted by excited atoms. Excitation can be brought about by either thermal or electrical energy or both. This is essentially a multi-element method, with a very wide concentration range.

Non-flame excitation sources used in AES are the *direct current arc* (DC arc) and the *high voltage spark* (HV spark). These were once frequently used but have been largely replaced by more effective excitation methods. They provided valuable qualitative and semi-quantitative results on metals in foodstuffs, soils and other materials.

Flame AES, though lacking some of the advantages of atomic absorption which has largely replaced it as method of choice for trace element analysis, is still widely used today, particularly for the determination of alkali metals. From the point of view of economy and simplicity, the method has some advantages over AAS. Simple and relatively inexpensive single-purpose flame photometers are available, though many AAS instruments also offer the AES as a built-in optional mode of operation.

Inductively Coupled Plasma Atomic Emission Spectrophotometry (ICP-AES)

ICP-AES uses a plasma source for atomisation–excitation and this has significantly improved detection limits, accuracy and precision of atomic emission procedures. Plasma flames are flamelike electrical discharges classified according to whether discharges occur between two electrodes, *direct current plasma* (DCP), or a radiofrequency discharge is used to transfer energy to a gas from an electrical power source radiofrequency, *inductively coupled plasma* (ICP). ICP-AES is now widely used in food analysis and many new developments in equipment are becoming available[56]. However, cost still restricts its use in many less well-equipped laboratories. While the technique has been found to allow simultaneous multi-element analysis with an accuracy in most cases approaching that of flame and electrothermal AAS, the technique requires, as has been noted by Jones[56], considerable care and a great deal of effort to obtain consistently reliable results. The same author believes that 'highly experienced analysts in trace element chemistry and ICP-AES spectrometry are essential' when analysing dietary components.

Among recent and most promising developments in this analytical technique have been the coupling of high performance liquid chromatography and mass spectrometry with plasma emission spectrometers. Both techniques have potential for greatly extending the range and sensitivity of trace metal analysis. ICP-MS is a particularly promising procedure[57]. However, their use is unlikely to become common in smaller laboratories because of the high costs and the great technical skills involved.

With regard to ICP-MS, a recent comment by two experienced British analysts is of interest[58]:

The analytical 'figures of merit' for trace element analysis using ICP-MS show that the technique has come of age and has good accuracy, precision and long-term stability for even the most difficult samples. This analytical performance, combined with the ease of daily routine set-up provided by careful interface design, makes ICP-MS worthy of serious consideration in most routine applications.

X-Ray Emission (XRE)

XRE requires an X-ray generating system, a sample compartment, a dispersion/detection system and a readout facility[59]. When samples are bombarded by high-energy atomic particles, X-rays or high energy electrons, X-ray fluorescence, (XRF) can be detected and measured. The electromagnetic radiation emitted presents a characteristic spectrum for each excited element. The intensity of the emission is a measure of the concentration of the excited element.

XRE and XRF provide the basis for a non-destructive method of simultaneous multi-element analysis of foods and other biological materials.

Particle-induced X-Ray Emission (PIXE)[60]

PIXE uses a beam from an accelerator or a cyclotron to induce emission of X-rays from samples. It is a multi-element analytical technique with a high level of accuracy and reliability which can operate with great speed and sensitivity. However, like the two other X-ray techniques described above, instrumentation is costly and requires considerable technical resources, normally out of the reach of the ordinary food scientist.

Neutron Activation Analysis (NAA)

NAA[61] requires the availability of a nuclear reactor for its operation. Samples (dried and sealed in quartz ampoules) are irradiated with a stream of thermal neutrons in a nuclear reactor. A proportion of the elements present in the samples is converted into radioisotopes. The type and energy of the radiation and the decay rate are specific identifying characteristics. The amount of the induced radioactivity is related to the quantity of the elements present in the sample.

NAA is one of the most powerful trace element analytical techniques available today[62]. It is also beyond the reach of most food laboratories. However, where there is access to an appropriate nuclear facility, under contract or a fee-paying arrangement, it is excellent practice to have analytical results obtained in the home laboratory checked by NAA as part of essential quality control procedures. This is particularly important as a check on the possibility of contamination of samples during analyses using laboratory equipment. In NAA the risk of contamination is very low since, prior to irradiation, manipulation of samples may be kept to a minimum. After irradiation, contamination is of no consequence since the added contaminants are not radioactive and thus are not detected.

Conclusion

An interesting comment is made by Versieck and Cornelis at the end of the chapter on analytical techniques in their text on trace elements referred to above[4]:

A bird's eye view of the tables with reference values given in Chapter 4 of this book, reveals that AAS is the leading technique in clinical laboratories. A very rough statistical treatment of the *circa* 130 data reported for the 15 elements shows that indeed about 51 per cent of all relevant values have been obtained with one or other form of AAS, with GFAAS the predominant method. NAA accounts for 27 per cent of the data, FLU for *circa* 6 percent, PIXE and XRF each for 3 per cent and ICP-AES, IDMS (isotope-dilution MS), potentiometric stripping analysis...., AES, flame emission spectrometry (FES) each between 1 and 2 percent. It may be anticipated that ICP-AES will, in the long run, take over a larger share of the workload.

What the authors say of the clinical laboratory holds true for the food analyst's laboratory also. In the following chapters, where individual metals are discussed and reference is made to appropriate methods of analysis for them, it will be seen that AAS is the principal analytical technique referred to in the literature cited.

REFERENCES

1. STEWART, K.K. (1979). Nutrient analysis of foods: State of the art for routine analysis, in *Nutrient Analysis Symp. Proc. Assoc. Off. Anal. Chem.* (Arlington, VA).
2. ALVAREZ, R., WOLF, W.R. and MERTZ, W. (1979). Biological reference materials certified for chromium content, in *Chromium in Nutrition and Metabolism*, eds SHAPCOTT, D. and HUBERT, J. (Elsevier, New York).
3. DYBCZYNSKI, R., VEGLIA, A. and SUSCHNY, O. (1980). *Report on the Intercomparison Run A-11 for the Determination of Inorganic Constituents of Milk Powder*. IAEA, Vienna.
4. VERSIECK, J. and CORNELIS, R. (1989). *Trace Elements in Human Plasma or Serum*, 65 (CRC Press, Boca Raton, Florida).

5. STEWART, K.K. and WHITAKER J.R. (1984). *Modern Methods of Food Analysis*, (Avi, Westport, Conn.).
6. *CRC CRITICAL REVIEWS IN ANALYTICAL CHEMISTRY.* e.g. Vol. 2, 1981. ed. Campbell, B. (CRC Press, Boca Raton, Florida).
7. MARCZENKO, Z. (1986). *Separation and Spectrophotometric Determination of Elements* (Horwood, Chichester, UK).
8. VARMA, A. (1984). *CRC HANDBOOK OF ATOMIC ABSORPTION SPECTROPHOTOMETRY* (CRC Press, Boca Raton, Florida).
9. VERSIECK, J. and CORNELIS, R. (1989). *Trace Elements in Human Plasma or Serum* (CRC Press, Boca Raton, Florida).
10. SCHMITT, Y. (1987). *J. Trace Elem. Electrolytes Health Dis.*, 1, 107–14.; KUMPULAINEN, J. and PAAKI, M. (1987). *Z. Anal. Chem.*, 326, 684–9.
11. FELL, G.S. (1983). *J. Inher. Metab. Dis.*, 6, Suppl. 1, 5–8.
12. LENTO, H.G. (1984). *Sample Preparation and its Role in Nutritional Analysis*, in STEWART, K.K. and WHITAKER, J.R. (1984). *Modern Methods of Food Analysis* (Avi, Westport, Conn.).
13. FISHER, R.A. and YATES, F. (1937). *Statistical Tables for Biological Medical and Agricultural Research* (Oliver and Boyd, Edinburgh).
14. SCHMITT, Y. (1987). *J. Trace Elem. Electrolytes Health Dis.*, 1, 107–14.
15. KUMPULAINEN, J. AND PAAKI, M. (1987). *Fresenius Z. Anal. Chem.*, 326, 684–9.
16. BENZO, Z., SCHORIN, H. and VELOZA, M. (1986). Simultaneous quantitative determination of manganese, iron, copper and zinc by atomic absorption spectrophotometry in tropical cereals, fruit and legume materials. *J. Food Sc.*, 51, 222–4.
17. KATZ, S.A. (1985). Metals in atmospheric dust. *ICPR* 13–19.
18. CORNELIS, R. and SCHUTYSER, P. (1984). Analytical problems related to the determination in body fluids. *Contr. Nephrol.*, 38, 1–11.
19. SANSONI, B. and IYENGAR, G.V. (1980). Contamination control in trace element analysis, in *Elemental Analysis of Biological Materials: Current Problems and Techniques with Special Reference to Trace Elements*, 57–71, *Tech. Rep. Ser.* 197. (IAEA, Vienna).
20. MOODY, J.R. and BEARY, E.S. (1982). Purified reagents for trace metal analysis. *Talanta*, 29, 1003–10.
21. VEILLON, C. (1986). Trace metal analysis of biological samples. *Anal. Chem.*, 58, 851A–66A.
22. SANSONI, B. and PANDAY, Y.K. (1983). Ashing in Trace Element Analysis of Biological Materials, in *Analytical Techniques for Heavy Metals in Biological Fluid*, ed. FACHETTI, S. (Elsevier, Amsterdam).
23. GORSUCH, T.T. (1959). Destruction of organic matter. *Analyst*, 84, 135–73.
24. ANALYTICAL METHODS COMMITTEE (1960). Methods for the destruction of organic matter. *Analyst*, 85, 643–56.
25. BUNKER, V.W., STANFIELD, M.F. and CLAYTON, B.E. (1986). *Hum. Nutr.: Appl. Nutr.*, 40A, 323–30.; ANDERSON, R.A. and KOZLOVSKY, A.S. (1985). *Am. J. Clin. Nutr.*, 41, 1177–83.
26. MAHER, W.A. (1982). Fluorimetric determination of selenium in some marine materials. *Talanta*, 29, 1117–18.
27. ROESCH, K. (1987). *The nutritional significance of certain trace elements in*

selected Australian diets. M. App. Sc. thesis, Queensland University of Technology.

28. SALMELA, S., VOURI, E. and KILPIO, J.O. (1981). The effect of washing procedures on trace element content of human hair. *Anal. Chim. Acta.*, 125, 131–7.; DEANTONIO, S.M., KATZ, S.A., SCHEINER, D.M. and WOOD, J.D. (1982). Anatomically-related variations in trace metal concentrations in hair. *Clin. Chem.*, 28, 2411–3.

29. FARRE, R., LAGANDA, M.J. and MONTORO, R. (1986). Atomic absorption spectrophotometric determination of chromium in foods. *J. Assoc. Off. Anal. Chem.*, 69, 876–9.

30. HALLS, D.J., MOHL, C. and STOEPPLER, M. (1987). Application of rapid furnace programmes in atomic absorption spectrometry to the determination of lead, chromium and copper in digests of plant materials. *Analyst*, 112, 185–9.

31. TINGGI, U. and MAHER, W., (1986). Determination of trace elements in biological tissues by aluminium block digestion. *Microchem. J.*, 33, 304–8.

32. MAHER, W.A. (1987). Decomposition of marine biological materials for the determination of selenium by fluorescence spectrometry. *Microchem. J.*, 35, 125–9.

33. CAPELLI, R., MINGANTI, V., SEMINO, G. and BERTARINI, W. (1986). The presence of mercury and selenium in human placenta. *Sci. Total Environ.*, 48, 69–79.

34. HANSSON, L., PETTERSSON, J. and OLIN, A. (1987). A comparison of two digestion procedures for the determination of selenium in biological material. *Talenta*, 34, 829–33.

35. WENLOCK, R.W., BUSS, D.H. and DIXON, E.J. (1979). Trace Nutrients 2: Manganese in British foods. *Br. J. Nutr.*, 41, 253–61.; LAWLER, M.R. and KLEVAY, L.M. (1984). Copper and zinc in selected foods. *J. Am. Diet. Assoc.*, 84, 1028–30.

36. ARAFAT, N.M. and GLOOSCHENKO, W.A. (1981). Method for the simultaneous determination of arsenic, aluminium, iron, zinc, chromium and copper in plant tissue without the use of perchloric acid. *Analyst*, 106, 1174–8.; BENZO, Z., SCHORIN, H. and VELOSA, M. (1986). Simultaneous quantitative determination of manganese, iron, copper and zinc by atomic absorption spectrophotometry in tropical cereals, fruit and legume materials. *J. Food Sci.*, 51, 222–4.

37. BUNKER, V.W., LAWSON, M.S., STANSFIELD, M.F. and CLAYTON, B.E. (1988). Selenium balance studies in apparently healthy and housebound elderly people. *Br. J. Nutr.*, 59, 171–80.

38. NEVE, J., HANOCQ, M., MOLLE, L. and LEFEBVRE, G. (1982). Studies of some systematic errors during the determination of total selenium and some of its tonic species in biological materials. *Analyst*, 107, 934–41.

39. ANALYTICAL METHODS COMMITTEE (1960). Methods for the destruction of organic matter. *Analyst*, 85, 643–56.

40. CROSBY, H.T. (1977). Determination of metals in food. *Analyst*, 102, 1213–68.

41. ISAAC, R.A. and JOHNSON, W.C. (1975). Collaborative study of wet and dry ashing procedures for elemental analysis of plant tissue. *J. Assoc. Off. Anal. Chem.*, 58, 436–40.

42. GORSUCH, T.T. (1962). Losses of trace elements during oxidation of organic materials. *Analyst*, 87, 112–24.
43. JONES, G.B., BUCKLEY, R.A. and CHANDLER, C.S. (1975). The volatility of chromium from brewers yeast during assay. *Anal. Chem. Acta.*, 30, 389–92.
44. CHAO, S.S. and PICKETT, E.E. (1980). Trace chromium determination by furnace atomic absorption spectrometry. *Anal. Chem.*, 52, 335–9.
45. HALLS, D.J., MOHL, C. and STOEPPLER, M. (1987). Application of rapid furnace programmes in atomic absorption spectrometry to the determination of lead, chromium and copper in digests of plant materials. *Analyst*, 112, 185–9.; CARY, E.E. (1985). Electrothermal atomic absorption spectroscopic determination of chromium in plant tissues: interlaboratory standards. *J. Assoc. Off. Anal. Chem.*, 68, 495–8.
46. IHNAT, M. (1984). Atomic absorption and plasma atomic emission spectrometry, in *Modern Methods of Food Analysis*, eds STEWART, K.K. and WHITAKER, J., 129–66 (Avi, Westport, Conn.).
47. ALVAREZ, R. (1984). NBS Standard Reference Materials for food analysis, in *Modern Methods of Food Analysis*, eds STEWART, K.K. and WHITAKER, J.R., (Avi, Westport, Conn.).
48. MACLEOD. A.J. (1973). *Instrumental Methods of Food Analysis* (Elek Science, London).
49. MARCZENKO, Z. (1986). *Separation and Spectrophotometric Determination of Elements* (Horwood, Chichester).
50. SANDELL, E.B. (1959). *Colorimetric Determination of Traces of Metals*, 3rd edn (Interscience, London).
51. CASSIDY, R.M. and ELCHUK, S. (1981). Trace enrichment methods in HPLC. *J. Chromatog. Sci.*, 19, 503–10.
52. KOH, T.S. and BENSON, T.H. (1983). Critical reappraisal of fluorimetric method for determination of selenium in biological materials. *J. Assoc. Off. Anal. Chem.*, 66, 918–26.
53. CANTLE, J.E. (ed.) (1982). *Techniques and Instrumentation in Analytical Chemistry*, Vol. 5, *Atomic Absorption Spectrophotometry* (Elsevier, Amsterdam).
54. EVANS, W.H., DELLAR, D., LUCAS B.E., JACKSON, F.J. and READ, J.I. (1980). Observations on the determination of total copper, iron, manganese and zinc in foodstuffs by flame atomic absorption spectrophotometry. *Analyst*, 105, 529–41.
55. FERNANDEZ, F.J., MYERS, S.A. and SLAVIN, S. (1980). Background corrections in atomic absorption utilizing the Zeeman effect. *Anal. Chem.*, 52, 741–6.
56. JONES, J.W. (1988). Radio frequency inductively coupled plasma. *J. Res. Nat. Bur. Stds.*, 93, 358–61.
57. HOUK, R.S. (1986). Mass spectrometry of inductively coupled plasmas. *Anal. Chem.*, 58, 97A–106A.
58. FULFORD, J.E. and GALE, B.C. (1987). Trace element analysis by inductively coupled plasma–mass spectrometry. *Anal. Proc. Roy. Soc. Chem.*, 24, 10–12.
59. VAN GRIEKEN, R., MARKOWICZ, A. and TOROK, S. (1986). Energy-dispersive X-ray spectrometry. *Fresenius Z. Anal. Chem.*, 324, 825–36.

60. JOHANSSON, S.A.E. (1986). Proton-induced X-ray emission (PIXE) spectrometry. *Fresenius Z. Anal. Chem.*, 324, 635–45.
61. ERDTMANN, G. and PETRI, H. (1986). Nuclear activation analysis: fundamentals and techniques, in *Treatise on Analytical Chemistry, Part 1. Theory and Practice*, Vol. 14, Section K. *Nuclear Activation and Radioisotopic Methods of Analysis*, eds ELVING, P.J., KIRVAN, V. and KOLTHOFF, I.M., 419–78 (Wiley, New York).
62. HISLOP, J.S. (1980). Choice of analytical methods, in *Element Analytical Chemistry in Medicine and Biology*, eds BRATTER, P. and SCHRAMEL, P., 747–67 (De Gruyter, Berlin).

PART II

The Individual Metals

CHAPTER 5

LEAD

INTRODUCTION

In the sixteenth century, Georgius Agricola, in his *De Re Metallica*[1], described lead as a 'pestilential and noxious metal'. Agricola was not the first, and certainly not the last, authority to point to the hazards associated with the metal. As early as about 400BC, Hippocrates, the 'Father of Medicine', described a disease which he called *saturnism*, with symptoms ranging from colic to delirium and paralysis, which occurred in men who worked with lead. Four centuries later, a Roman observer noted that workers in shipyards, who painted a white pigment made from lead as a preservative on the hulls of vessels, wore cloth masks to protect themselves against lead poisoning, or, as he knew it, *plumbism*.

In this century, in which we pride ourselves on the advances that have been made in occupational health and safety, it can be with no great pride that we see that industrial lead poisoning still occurs. A review of lead and health published in 1978[2], reads:

Lead has a wide range of applications, and its production and use result in contamination of the environment, including food and drinking water. Geochemical studies indicate that the majority of lead in ecosystems originated from industrial operations, and that human lead intake has increased 100-fold above the 'natural level'. Prehistoric human skeletons contain about two orders of magnitude less lead than present-day samples...the highest occupational lead exposures occur in lead smelters and storage battery plants, but several other industrial working operations may result in high lead levels. As much as 1 per cent of the working population may have a significantly increased lead absorption with possible adverse effects.

Hippocrates might not have understood the technical language, but he would have appreciated the message of this review: lead continues to be a major health problem in the world. It is for this reason that, as has been mentioned in an earlier chapter on food legislation, lead holds a prominent place in the health and food laws of most countries. Indeed, it is probably the metal which has had more official consideration and about which more has been written than any other inorganic contaminant of food and the environment. However, as we will see, mercury and cadmium are not very far behind lead in this regard.

CHEMICAL AND PHYSICAL PROPERTIES

Lead is element number 82, in group IVB of the Periodic Table, with atomic weight of 207.19. It is one of the heavier of the elements, with a density of 11.4. It is soft and bendable and can be beaten flat with a hammer and cut with a knife. When first cut, its surface is bright and mirror-like, but this soon tarnishes on exposure in moist air and a surface film of basic lead carbonate forms. A similar protective film forms when the metal is placed in 'hard' water containing carbonates.

Lead melts at a relatively low temperature of 327° C. Its boiling point is 1725° C. The metal is a poor conductor of heat and of electricity. Oxidation states are 0, + 2 and + 4. In organic compounds lead is usually in state + 2. Most salts of lead (II), lead oxides and lead sulphide are only slightly soluble in water, with the exception of lead acetate, lead chlorate and, to a lesser extent, lead chloride. Lead forms a number of compounds of technological importance. When heated in air, a yellow powdery monoxide, PbO (litharge), is formed. On further heating the monoxide is raised to a higher oxidation level, producing trilead tetroxide Pb_3O_4 (red lead).

White lead or basic lead carbonate, $Pb_3(OH)_2(CO_3)_2$, is prepared commercially by the action of air, carbon dioxide and acetic acid vapour on metallic lead. Another coloured compound is yellow lead chromate, $PbCrO_4$.

Some of the organic compounds of lead are of great economic importance. Tetraethyl lead, $Pb(C_2H_5)_4$, is liquid at normal temperatures with a boiling point of 200° C, and has antipercussive properties. It is prepared by the reaction of ethyl chloride, C_2H_5Cl, with sodium–lead alloy. Tetramethyl lead, $Pb(CH_3)_4$, is also a liquid, with a boiling point of 110° C. Both these organolead compounds decompose at or just below their boiling points.

Lead has an ability to form alloys with other metals. Some of these, such as solder which is made with tin, are of considerable economic importance. Formerly, pewter contained 20 per cent, or more, lead alloyed with tin, but modern pewters contain little, if any, lead.

PRODUCTION AND USES

Lead is found, at least in small amounts, almost everywhere in the world. Soils normally contain between 2 and 200 mg/kg. The metal occurs combined with other elements as salts. It is usually associated with zinc, iron, cadmium and silver. Economically workable ore bodies occur in many parts of the world. The major lead-mining countries are the USA, USSR, Australia, Canada, Peru, Mexico, China, Yugoslavia and Bulgaria. The most common ores are galena (lead sulphide), cerussite (lead carbonate) and anglesite (lead sulphate).

Lead is used on a very wide and increasing scale in the modern world. Data published by the International Lead and Zinc Study Group (ILZSG)[3] show that lead production in the Western world totalled 2.131 million tonnes in the first half of 1988, with 1.255 million tonnes of this smelted from ore and the remainder recovered from scrap. In that time, use of lead rose by 3.2 per cent to 2.109 million tonnes, compared to the same period in 1987. It is estimated that total world production of lead will be about 6 million tonnes by the end of the century, with an increasing amount of this recycled from scrap.

The ease with which lead may be extracted from its ores probably accounted for its early exploitation by man. When, for example, galena is roasted on a charcoal fire it is partly oxidised at a low temperature and then proceeds, more or less by itself, to complete the conversion to a metallic state. The lead oxide reacts with the unconverted lead sulphide and releases the free metal. This is the system on which the original ore hearth furnace was based, a method used by the Romans 2000 years ago, and still in limited use until a short time ago. Today it has been replaced by the blast-furnace, in which reduction is achieved by carbon monoxide or producer gas made from coke. The technique with sulphide ore (3–8 per cent lead) includes concentration (to 50–70 per cent lead), sintering, in which the lead is oxidised, followed by reduction and then, finally, refining of the metal.

The uses of lead fall into two main groups: first as the metal and second in chemical compounds[4]. In Britain, about two-thirds of the total consumption comes into the first group and is related to the metal's high resistance to corrosion.

The largest use of lead throughout the world today is in lead batteries — electrical accumulators for cars, electric vehicles, emergency lighting and other such uses. This accounts for about 40 per cent of total world consumption. It has been estimated that almost all the lead used in batteries is recycled from scrap, and the turnover time is about four years.

Lead-sheathed cables form another major use. Formally, a great deal of the metal was used in the manufacture of lead pipes but, with the introduction of plastic piping, this use decreases year by year. Similarly, lead sheet is used less today than in former times for roofing. It is still extensively employed in chemical plant for handling sulphuric acid.

Lead finds a variety of other uses. It has traditionally been used for making bullets and other projectiles for guns and is still employed in great quantities for this purpose. In France alone, for instance, some 12 000 tonnes of lead were used in the manufacture of shot for gun cartridges in 1974[5].

Lead solders of various composition are used extensively in a number of industries, especially in plumbing and electrical work. They are of use for joining seams in cans used to contain food. In the motor industry, lead and lead solders are used in car body manufacture and in bearings. In fact, more than 50 per cent of the total world production of lead is for the motor industry, either in engine and body manufacture and accumulators or in fuel additives.

Lead alloys are used in considerable quantity as printing metals, though normally the same metal is used over and over again as type-face is melted down and recast. This use is, however, decreasing as modern computer technology takes over from the old lead font.

The use of lead compounds, both organic and inorganic, is extensive and continuously increasing. A certain amount of lead monoxide is still used today to manufacture white paint, but it is now largely restricted to paints used for priming external work. Formally, much greater quantities of white lead were used in paint manufacture. Many old buildings still have layers of this toxic pigment on their walls and woodwork.

The widespread use of white lead paint on wooden buildings and furniture has left a legacy of lead poisoning in several parts of the world. In Queensland, Australia, the toxicity of white lead paint and its dangers, especially to children, was recognised as long ago as the second decade of this century[6].

A lead pigment still used in large quantities is red lead. Along with the more recently developed calcium plumbate, it is used as a rust inhibitor for iron and steel. An idea of the quantity of such paints used in protecting steel and iron structures is given by the fact that 91 tonnes are used each year in the continuous painting that is carried out on the Sydney Harbour Bridge in Australia. Exposure to this quantity of lead paint has its hazards. A recent investigation found that of the 80 painters who work on the bridge, 18 had significant lead intoxication[7].

A major use of lead salts is in glazing of ceramics. Lead salts are also used extensively in glass manufacture, both for high-quality crystal and for television tubes and fluorescent lights. Though, as has been mentioned, the replacement of lead pipes and cisterns by plastics has led to a reduction in the use of the metal in plumbing, plastics such as polyvinyl chloride (PVC) require lead salts as stabilisers.

There are many other minor uses of lead which between them use about 3000 tonnes of the metal annually in Britain. They include driers in paint and printing inks, agricultural fungicides and insecticides, and antifouling compounds for hulls of ships.

Another major use of lead is in the manufacture of petroleum additives. However, this use is decreasing as leaded petrol is phased out in many countries. In the early 1970s, before the use of lead-free petrol became widespread, in the UK alone approximately 40000 tonnes of metallic lead were converted into tetraethyl and tetramethyl lead as a fuel additive. There was growing, and very vocal, concern in many countries at the enormous environmental burden that was resulting from the use of leaded petrol. A report on the problem commissioned by the Australian Academy of Science[8], noted that lead in motor-exhausts had become the major source of emission of the metal into the atmosphere, especially in cities. At that time 8000 tonnes of lead were added each year to petrol in Australia. Of this it was estimated that about 5600 tonnes were emitted directly in exhaust fumes and up to 600 tonnes by burning of sump oil and smelting of steel automobile scrap. In contrast, emission from primary and secondary lead smelters, and from fossil fuel-burning power stations, was believed to total a maximum of 1500 tonnes.

LEAD IN THE HUMAN BODY

Lead was listed by Schroeder and Nason among the 'abnormal trace elements of interest in the human body and blood'[9]. It is certainly of interest, but is hardly abnormal, because lead is present in practically every organ and tissue of the body. The total amount present varies with age, occupation and even race. It has been estimated that western 'Reference Man', a 70-kg male who has not been exposed to excessive amounts of environmental lead, contains between 100 and 400 mg of lead with an average of 120 mg, or 1.7 μg/g of tissue. Of this, 1.4 mg is in blood and more than 100 mg, or some 92 per cent of the total, in bone. In Britain the concentration of lead in bones of men and women over 16 years of age ranges from 9 to 34 mg/kg. Liver contains about 1 mg/kg and kidney somewhat less. Similar levels have been observed in Japan and the USA. In Britain the level ranges between 0.02 and 0.8 mg/kg[10].

We probably begin life with a small body store of lead. Transfer across the placenta from the mother to the foetus has been shown to occur. The lead content of the body increases with age, as our exposure to environmental lead is prolonged. This lifelong build-up is mainly in bone, which acts as a reservoir for the metal. It is not totally and permanently fixed there, but under certain conditions of stress may be released into the bloodstream.

After birth we absorb our body's lead either from our food and drink or from the air. Studies by Kehoe and others[11] have indicated that about 10 per cent of ingested lead is absorbed from the gastro-intestinal tract of adults. There are indications that the amount absorbed by children may be higher. As much as 53 per cent of dietary lead was absorbed by a group of children ranging in age from 3 months to 8 years[12]. Newborn animals have also been shown to have a high absorption rate.

Several dietary factors can affect the level of absorption. A low body-calcium status results in an increased absorption of lead. Uptake of both calcium and lead is increased by vitamin D. Iron deficiency may also promote lead absorption, as will fasting. A diet high in carbohydrates but lacking in protein can have a similar effect.

Once absorbed into the bloodstream, lead is transported round the body like most other heavy metals, by being attached to blood cells and constituents of the plasma. It is distributed to form an exchangeable compartment (blood and soft tissue) and a storage compartment (bone). Lead in blood is bound mainly to the erythrocytes, where its concentration is about 16 times higher than in plasma. Some of the lead absorbed is transported to the brain. Metallic lead, however, does not accumulate there to any great extent. Tetraethyl lead, on the other hand, appears to be retained preferentially in brain tissue, where it is partially broken down to form the triethyl derivative.

Lead is eliminated from the body in faeces, as well as in urine and in sweat. About 90 per cent of ingested lead is lost in faeces. Of the 10 per cent that is actually absorbed from the diet, about three-quarters will eventually leave the body in urine. Other excretory pathways are gastro-intestinal secretions, hair, nails and sweat. Schroeder believed that the latter pathway is of considerable importance and noted that the 'sense of well-being resulting from sauna baths, vigorous exercise or exposure to the hot and low lead environment of the tropics is the result of negative lead balances induced by sweating'.

Lead finds its way out of the body in milk. Levels of as high as 0.12 mg/litre have been found in human milk in Japan[13]. Figures published by Dabeka and McKenzie[14] for Canada range from <0.05 to 15.8 μg/litre. These lower figures may reflect the absence of gross environmental pollution in Canada compared to Japan. However, they could also be the result of more accurate analytical procedures introduced in the two decades that separate the studies. There is considerable evidence, however, of increased levels of lead in human as well as cows' milk in areas where environmental lead is abnormally high. While it is theoretically possible that human infants could be poisoned by lead in mother's milk, this has not been shown to occur in practice. However, poisoning has been produced in suckling rats and mice by exposing their mothers to lead.

There is a similarity between the metabolism of lead and of calcium. Both metals are found in bone crystal structure, which consists mainly of calcium phosphate. It was once believed that, with time, lead became buried deeper and deeper within the bone structure and was held there in a permanent state. However, there is now evidence that such 'burying' takes place only over a very long time and that a potentially dangerous pool of available lead persists for many months after exposure to high levels of lead. It is also possible that even 'fixed' lead may be mobilised from bone under certain conditions, such as shock and illness.

Levels of lead detected in the bones of ancient skeletons uncovered in archaeological digs in different parts of the world suggest that modern man accumulates more of the metal than did his predecessors. While lead in the skeletons of 258 twentieth-century Americans was found to vary from 7.5 to 195 mg/kg, with an average of 100 mg/kg, levels in skeletons of Peruvian Indians dating from about AD 1200 were less than 5 mg/kg. Another study of third-century Polish skeletons, found that these early Europeans had between 5 and 10 mg/kg, in contrast to modern Poles whose skeletons contain about 12 mg/kg[15].

Lead in hair has been used as an indicator of exposure to heavy metal contamination and to assess levels of dietary uptake. However, there is considerable lack of agreement about the reliability of such assessments. Hair of children, for instance, has been used in an attempt to examine the level of exposure to lead in the USA over the past century[16]. Locks of hair preserved in lockets from children who lived between 1871 and 1923, as well as fresh samples of modern hair, were analysed. The 'antique' hair had a mean of 164 mg/kg of lead compared to 16 mg/kg in modern hair. This reduction, it was claimed, resulted from changes in the American way of life. These changes included the fact that drinking water was no longer collected from metal roofs and poorly glazed earthenware vessels were no longer used for holding food and beverages. However, others[17] have shown that the assessment of lead contamination from hair levels is not a straightforward matter. Age and sex, as well as the presence of other metals in the diet, can affect accumulation. The simultaneous presence of cadmium, for instance, can enhance the uptake of lead considerably. A survey of hair-lead carried out among students and others in Oxford, England, showed[18] that there was considerable variation in levels even in samples from those who lived in the same area and had, apparently, a similar way of life.

BIOLOGICAL EFFECTS OF LEAD

The symptoms of acute lead poisoning, following ingestion of large quantities of the metal, are well known. Less is known about chronic effects which may occur after the accumulation of lead in the body over a long period of time. The major effects are related to four organ systems: haemopoietic, nervous, gastro-intestinal and renal. Acute lead poisoning usually manifests itself in gastro-intestinal effects. Anorexia, dyspepsia and constipation may be followed by an attack of colic with intense paroxysmal abdominal pain. This is the 'dry gripes' of the 'Devonshire colic', formerly known to cider and spirit makers who used lead-lined vessels in their fermenting and distillation processes. At times the pain from lead colic is so severe that it can be taken for acute appendicitis.

Lead encephalopathy in adults is rare but it occurs more frequently in children. It has been observed in children with pica in the USA[19] and in African children living in the vicinity of a zinc–lead smelter[20].

Today, gross symptoms of lead poisoning are seldom met with except in those exposed to extreme occupational hazards. Attention is largely centred on subclinical levels of poisoning, the chronic condition that may follow on residence in a high-lead environment or ingestion over a long period of time of small amounts of the metal in food. One of the effects of such subclinical poisoning is interference with the pathway of haem biosynthesis. Mild anaemia is often observed among occupationally exposed workers. Though it is seldom noted today, the characteristic 'pallor' of lead workers was first described as long ago as 1831 by the French physician Laennec. This lead-induced anaemia is caused by the combined effect of the inhibition of haemoglobin synthesis and shortened life span of the circulating erythrocytes. Many stages in the pathway of haem synthesis are inhibited by lead. Delta-aminolevulinic acid dehydrase (ALA-D), which catalyses the formation of porphobilinogen from o-aminolevulinic acid (ALA), and haem synthetase (haem-S), which incorporates iron into protoporphyrin IX, are the enzymes most affected. However, other stages can also be blocked.

Measurement of the activity of ALA-D activity, as indicated by concentration of ALA in the blood, provides a sensitive clinical test of lead poisoning. It gives an indication of levels of lead uptake long before any obvious physical symptoms appear. This is of value in detecting incipient lead poisoning.

Apart from acute encephalopathy, milder symptoms of the effects of lead on the central nervous system occur. These include mental deterioration and hyperkinetic or aggressive behaviour. However, it is very difficult to establish a direct relationship between elevated levels of blood-lead and neurophysiological effects, especially if they are sub-clinical. Cerebral injury due to viral or other causes has been known to result in persistent pica, with a consequent increased intake of lead, thus making high blood-lead levels a consequence rather than a cause of mental deterioration.

Peripheral neuropathy was formerly frequently observed among workers exposed to lead and in others whose intake of the metal in food and drink was excessive. Lead palsy resulting in wrist or foot drop, in which there is actual paralysis of the muscles of the hands and feet, is now seldom observed. During prohibition days of the 1930s in the USA, cases of ankle drop were reported among dancers in cabarets where illicit, lead-contaminated 'bath-tub' gin was consumed.

The cause of this palsy has been extensively studied. It is related to muscular fatigue. Muscle and nerve action depend on a balanced flow of calcium across cell membranes. Lead interferes with this movement by forming soluble lead lactate by combining with lactic acid produced during muscle metabolism. The lead lactate readily penetrates into muscle and nerve cells and there it is once more changed, by combining with phosphate, into an insoluble form. The lead phosphate settles like a barrier on the surface of cells and blocks free passage of calcium, resulting in neuromuscular effects.

There is evidence that chronic kidney disease may result from prolonged exposure to even small, regular intakes of lead. A WHO report[21] concluded that prolonged exposure to lead with blood levels above 70 μg/100 ml may result in chronic irreversible nephropathy. A report from Australia[22] supports this conclusion. This notes that the death rate from chronic nephritis in Queensland is far higher than elsewhere in the subcontinent. It is suggested that the cause of the high levels of nephritis deaths was childhood lead poisoning due to drinking water collected from roofs which had been painted with lead-based paints. White painted, wooden houses, each with their rainwater collecting tank, are characteristic of this subtropical state. The climate contributes to rapid deterioration and flaking of the paint and this can lead to contamination of tank water. Of 401 cases of childhood lead poisoning, follow-up studies showed that 165 died under the age of 40, and 101 of these deaths were attributed to kidney failure.

ALKYL LEAD

Over the past decade there has been a considerable reduction in the use of lead additives in petrol throughout the world. The need for measures to control and reduce the lead burden of the general community from all sources has been widely recognised. Prominent among the steps taken to achieve these results has been legislation to restrict and eventually totally exclude the use of alkyl lead in motor fuels. In Europe the EEC blood-lead directive[23], as well as national measures in individual members of the Community, have had a significant effect. Several reports show that there is a continuing decrease in blood-lead levels in Europe which is attributable to the reduction in permissible levels of lead in motor fuel[24]. This reduction is also reflected in an observed decrease in levels of lead contamination of foods exposed for sale in stores close to busy highways.

Unfortunately, tetraethyl lead (TEL) and other alkyl lead compounds are still widely used in motor fuels in many countries. TEL is an acutely toxic compound — far more dangerous than metallic lead or its inorganic compounds — and the poisoning it causes is quite different from that due to inorganic lead. It is very volatile and its absorption is usually by inhalation, but the liquid is easily taken in by the skin as well as by the gastrointestinal tract. After absorption TEL is distributed to various tissues, particularly to the brain and other organs, decomposing into triethyl lead and minute amounts of inorganic lead.

It is unlikely that TEL contamination of food occurs in more than isolated cases, but accidental drinking of contaminated beverages has occurred. Most known cases of TEL poisoning have resulted from inhalation during cleaning of petroleum storage tanks or from the misuse of gasoline as a cleaning fluid in badly ventilated work-rooms. Unfortunately, voluntary sniffing of petrol fumes is practised, usually by particular socially-deprived groups, in many parts of the world. It is a serious problem among aboriginal youths in Australia[25] and is also practised in some American Indian groups[26].

The earliest symptom of alkyl lead intoxication is insomnia and the main organ affected is the nervous system. Poisoning is usually acute, developing into toxic psychosis, and may result in death. Increased blood and hair levels of lead are observed in chronic sniffers and the symptoms of lead encephalopathy are sometimes observed[27].

Most of the TEL in motor fuels is decomposed by combustion into inorganic lead compounds such as lead halides and oxides. However, very small amounts of organic lead do escape into the ambient air. A US Department of Health, Education and Welfare survey[28] found that TEL concentrations in the urban atmosphere were less than 10 per cent of the inorganic lead levels and were too low to affect health.

TEL in motor fuels contributes a very great deal of inorganic lead to the environment, as was noted earlier. High concentrations of lead occur in soils and herbage near highways[8]. There is controversy about the extent of lead contamination of foods grown or offered for sale in the vicinity of motor traffic[29]. A Danish study carried out before a reduction in levels of lead in petrol had been widely enforced, suggested that lead from car exhaust is dispersed over a wide area[30]. However, this view has not been supported by other studies which have found that, while contamination may be high in close proximity to busy highways, the level of lead on plants decreases with distance from the source of the exhaust fumes[31].

A striking confirmation of this finding was provided by a study carried out at the Rothamsted Agricultural Experimental Station, Hertfordshire, England[32]. Samples of topsoil collected regularly at the station since 1881 were analysed for lead. Though the station is not on the immediate verge of a busy highway, it does lie within 2 km of three major trunk roads and is 42 km from the centre of London. Its soil and herbage lead levels might

TABLE 11
Lead Content of Some Foods

Food	Mean (mg/kg)	Range (mg/kg)
Cereals	0·17	< 0·01–0·81
Meat and fish	0·17	< 0·01–0·70
Fruit (fresh)	0·12	< 0·01–0·76
Fruit (canned	0·40	0·04–10·0
Vegetables (fresh)	0·22	< 0·01–1·5
Vegetables (canned)	0·24	0·01–1·5
Milk	0·03	< 0·01–0·08

be expected to reflect changes in the background level of the metal in this industrialised and heavily populated part of England. The results showed that soil levels of lead had only increased slightly during this century. Herbage samples which were collected over a similar period of time also failed to show any noticeable increase in lead content. It is particularly significant that the results do not in any way reflect the introduction of organic lead antiknock petrol additives in the 1930s or the increase in motor traffic in later years.

LEAD INTAKE IN THE DIET

Lead is a normal ingredient of our diet. It is present in all foods and beverages, both as a natural component and an accidental additive picked up during processing. Levels of the metal normally found in a variety of food types are shown in Table 11. Levels of this order are consistently found in foods in most countries of the world.

Considerably higher levels of lead in food occur as a result of environmental pollution by industry and other causes. Such local surges in lead levels in food can have serious consequences for health. Health authorities and legislators seek to prevent such health problems by establishing maximum permissible levels for lead in food which is sold to the public. In addition, the World Health Organization has established potential tolerable weekly intakes (PTWI-values) for lead, and other toxic metals, as guidelines for controlling the problem[33]. The PTWI-value for lead for an adult is 3 mg (or 430 μg/day).

Different estimates have been made of the amount of lead we ingest daily, but actual intake depends to some extent on where we live, the quality and origins of the food and drink we consume, and other factors. As has been noted by Louekari and Salminen in a study of intakes of heavy metals in certain European countries[34], published results from different countries are not always easy to interpret, because of the widely different methods used to estimate contaminant intake in different studies. Consequently, it is seldom clear whether the variation of intake between countries is due to differences in food consumption or to heavier environmental contamination. However, though differences in actual intakes are reported, the authors also note that, in all countries capable of completing intake estimations, the average intake of lead is below the WHO PTWI-value.

Kehoe[35] calculated that the average American consumed between 0.15 and 0.35 mg/day in food, with about another 0.1 mg coming from water and other beverages and from atmospheric pollution. Figures for the UK suggest an intake of lead in food and drink of between 0.055 and 0.366 mg/day[36]. A 1976 WHO Task Group concluded that a fair estimate of daily intake of lead via food was 0.200–0.300 mg[37]. Somewhat higher figures were reported by Horiuchi for Japan, with a range of 0.239–0.318 mg each day consumed by average adults, while men doing 'hard labour' consumed an average of 0.455 mg[38].

An interesting study by a Japanese group[39] calculated lead exposure in 30 countries. It was assumed that lead concentrations in foods would be the same in all the countries. The study showed clearly that food consumption patterns have a significant effect on dietary intake of lead, depending on the total amounts of food ingested and the relative importance of particular food groups in the average diet. As we will see below, this latter consideration is of major significance when it comes to lead and other toxic metal intake by such vulnerable groups as infants and young children. The overall results of the study found that average lead intake from food ranged from 71 μg/day in India to 229 μg/day in the USA.

More detailed information on lead intake in three countries, Finland, West Germany and Japan, is given in the Finnish report by Louekari and Salminen referred to above[34]. These authors found that mean lead intake by adults in West Germany was 180 μg/day and in Japan 165 μg/day, while in Finland adult intake at 70 μg/day, was 30–40 per cent less than in the other two countries. The authors commented that environmental contamination, leading to higher levels of lead in foods consumed, was obviously a major cause of these differences. For example, lead was 2.5–3 times lower in liver and 20–30 times lower in vegetables in Finland compared to the other two countries. However, the differences can also be partly explained by food consumption habits. The high intakes of fish and seaweed in Japan and of beer in West Germany, products relatively rich in lead compared to those consumed more commonly in Finland, also make a significant contribution to dietary lead intake. It is also of interest that Finns consume remarkably less vegetables than do Germans (87 g in Finland compared to 188 g in Germany each day; Japanese intake of vegetables is 299 g/day).

Louekari and Salminen note that it is probable that in some groups in Germany and Japan, the PTWI is approached and even exceeded, since the intake of heavy metals is known to vary very much within the population. This is an important conclusion, with implications for legislators and health authorities. An Australian study on health and environmental lead[40] indicates that there is also some cause for concern about levels of dietary lead intake in that country. This is, in fact, a surprising finding since, as the report notes, most of the food that Australians eat is grown in rural areas where the levels of lead in the soil and in the atmosphere are low. However, estimates of daily intake of lead are 'towards the top of the range and more than twice the values estimated for individuals in Great Britain'[8]. Intakes ranged from 0.105 mg for an infant to 0.345 mg for an adult woman and 0.494 mg for a man. The top male intake exceeds the WHO PWTI-value by 0.46 mg, a significant figure. However, it should be noted that the Australian results are based on calculations from hypothetical diets, while those in several other studies are based on results obtained by laboratory analyses of duplicates of foods consumed by a number of individuals. It is possible that the use of hypothetical diets leads to an over-estimation of quantities of food consumed and thus gives an inflated figure for lead intakes.

As was noted in the Finnish study, lead concentrations in individual items in the diet, and the amounts of a particular foodstuff consumed, can have a significant effect on overall dietary intake of lead. Thus, Germans drink three times as much beer and take about 10 times as much lead from the beverage as do the Finns. In fact Germans consume 23.3 μg of lead in 489 g of alcoholic beverages, mainly beer, each day compared to the Finns' 3.5 μg in 180 g. In contrast, the Finns drink 711 g of milk (containing 8.8 μg of lead) and the Germans 329 g of milk (with 3.4 μg of the metal). Australians show the same effect of high beer–higher lead intake as do the Germans. The Australian Academy of Science study found that the highest individual source of lead in the adult male diet was beer, providing 14 per cent of the total. Plant foods, including vegetables, fruit and fruit juices, accounted for 50–70 per cent of the total lead intake.

The significance of the type of cooking utensil used to prepare meals with regard to dietary lead intake should not be overlooked. It has been shown, for instance, that a meal cooked in an aluminium or stainless steel saucepan contained a lead content of 0.08 mg, but when a tinned copper pan was used this increased to 0.35 mg[41].

ABSORPTION OF LEAD FROM FOOD

Only about 10 per cent of ingested lead will actually be absorbed from the alimentary canal in an adult. There are, however, differences between absorption levels in different populations and individuals. A number of factors affect absorption. It will depend on whether a person is fasting or not, on the chemical form of the lead in food, as well as on other metals present, and on the fat content of the diet.

Lead absorption in children is believed to be higher than in adults. Average values of 25–53 per cent have been observed in children from 2 weeks to 8 years of age[42]. However, while a wide variation between individuals is recognised, it is generally believed that on average about 40 per cent of lead ingested by children is absorbed from the gut. This higher intake by children compared to adults is of considerable significance. It is the reason why health authorities and food legislators in many countries pay special attention to lead in foods for infants and children, and accounts for the stringent restrictions placed on maximum permitted levels in such foods.

Given that, on average, absorption of lead from food is 10 per cent in adults and 40 per cent in children, and that average intake of lead is below the PTWI, it can be estimated that between 20 and 30 μg of lead will enter the bloodstream of an adult each day. In the case of persons with no known high exposure, blood-lead levels, resulting mainly from ingested food, are fairly uniform throughout the world, ranging from between 10 and 35 μg/100 ml on average[43].

LEAD IN DIFFERENT TYPES AND GROUPS OF FOODS

Infant Formula

The level of lead ingested by infants, from whatever source, has been a cause of considerable concern for many years. In spite of improvements in food processing practices and a reduction in lead levels in petrol and consequent decrease in emission of the metal in car exhausts, there are grounds for believing that more needs to be done to decrease lead consumption in infancy. Recent studies have indicated that blood-lead levels previously accepted as normal are associated with learning defects[44]. There is also evidence that the developing brains of infants and growing children are highly susceptible to lead toxicity[45]. The major source of lead appears to be the diet, and absorption and retention are higher for infants and children than for adults[46].

A recent Canadian study[47] found that dietary intake of lead by infants averaged 16.5 μg/day (2.4 μg/kg/day). Of this total intake, 2.24 μg, or 0.35 μg/kg body weight, was provided by water used to dilute the formula. Lead intake increased with age, from 12.3 μg/day for the first month of life, to 19.7 μg/day at the end of the first year, but, in terms of body weight, this actually represents a decrease from 3.1 to 1.9 μg/kg/day. These figures were lower than those reported 8–10 years earlier. The reductions were attributed by the authors of the report both to improvements in Canadian canning procedures and to lower detection limits of the analytical methods.

A major source of lead in these Canadian infant diets was ready-to-use formula in lead-soldered cans which contained on average 46.2 ng/g of lead. In contrast, ready-to-use formula in lead-free cans contained only 1.7 ng/g, and in glass containers 2.5 ng/g of the metal. Evaporated milk also had significantly high levels of lead, with a range of 27–106 ng/g.

The average intake of lead in food and water by the infants was only slightly less than the provisional tolerable daily intake (PTDI) of 3.5 μg/kg/day. The authors of the report note that it is quite possible that individual infants, depending on the type of formula they receive and the quality of the water used in its preparation, ingested more lead than the PTDI.

A British study[48] found, like the Canadians, that water used in formula preparation could have a major effect on levels of lead ingested by infants. In the city of Glasgow, where lead pipes and solder are still found in domestic water systems, levels of lead in tap water are often above 100 μg/litre. This is twice the maximum allowable concentration (MAC) of 50 μg/litre established by the European Economic Community's Directive on drinking water quality[49] and ten times the average found in most households in England[50]. Infants fed on formula made up with this water were found to have lead intakes of up to 3.4 mg/week, higher than the PTWI of 3.0 mg/week for *adults*. About one-quarter of all the infants had intakes above 1.0 mg/week. A finding which caused particular concern was that several mothers use water from the hot tap to make up the infants' formula, which resulted in a higher level of lead than if the cold water tap had been used.

Meats

Meat, fish and other animal products have been implicated in a number of serious cases of poisoning due to the transmission of metals from the environment to man. There have been reports of fish and meat that were dangerously contaminated with mercury. High levels of cadmium have also been found in free-swimming fish as well as in shellfish. However, though lead is accumulated by oysters and crustaceans in polluted waters, there is no strong evidence that fish or indeed any other animal has been the vehicle by which lead poisoning is transmitted to man. However, offal, especially beef kidney, may in some circumstances have unacceptably high levels of lead. This can occur when cattle have grazed on sewage sludge-treated grass[51].

This may not be the case with game meat. In their review of food regulations relating to lead, the UK's Food Additives and Contaminants Committee in 1975 made an interesting, and perhaps, controversial recommendation with regard to game and game paté[52]. The Committee noted that 'it has been represented to us that it is virtually impossible for these foods to comply with the present statutory limit of 2.0 mg/kg. The possible presence of whole or discrete pieces of shot would make this limit unrealistic'. As a result of their deliberations, they therefore recommended that the permitted level of lead in game and game paté should be increased to 10 mg/kg 'not including discrete lead shot'.

Countries with a less traditional approach to game shooting might not be willing to accept the Committee's reasoning. They would quite rightly observe that the proposed standard would permit the legal sale of meat capable of causing lead poisoning if eaten in sufficient quantity. Game — the rabbits and hares, the ducks, geese, partridges and pheasants which for most of us are infrequent special items in the diet — are more often than not killed by lead pellets shot from a gun. The shot can lodge in the flesh and, if not noticed during preparation of the meal, may be consumed.

Even before they are brought down by a further shot, certain game birds, especially ducks and geese, may already be carrying a high load of lead.

Twenty years ago it was estimated that a staggering 15 billion individual lead shot were fired each season over the duck and geese hunting grounds of North America alone[53]. Only about one-fifth of these particles actually struck a bird. The remainder fell into the water or onto vegetation or the ground and there could be consumed by feeding birds and other animals. Lead-poisoned birds are frequently recovered by Wildlife and Fisheries agents of the USA and Canada. No doubt quite a portion of the birds killed by hunters and eaten by them, or by others who subsequently purchase them, contain high levels of lead.

A Jugoslavian study[54], which found that 41 per cent of mallard ducks killed during the 1979–82 hunting seasons, had more than 2 mg/kg of lead in their tissues and that many of these birds had lead shot in their gizzards, shows that the problem is not confined to North America alone. There is little information about the situation in the UK and other parts of the world, but there is sufficient evidence to show that game birds and animals could be a significant source of lead intake in these countries as well as in America. The danger to health is, of course, very restricted. Relatively few people will eat game regularly or in such large amounts that they will suffer lead poisoning as a result. However, it would be unwise to assume that in the case of infants and young children, game does not represent a better avoided hazard for health.

Water

Most natural waters contain about 5 μg of lead per litre. The WHO has recommended a maximum permissible limit of 50 μg/litre, which is the same as the EEC maximum allowable concentration. Normally municipal supplies will contain well below this limit but, as has been indicated above, this is not always the case. A survey carried out in Liverpool found that, of 47 samples from public water supplies, 22 had concentrations at or above the WHO recommended level. Comments have already been made about the situation in Glasgow. Such contamination is not confined to the UK, as results of investigations of municipal supplies in Boston and Seattle in the USA show[55].

Lead contamination of water supplies may be due to pollution of rivers, wells and other sources by industrial and municipal waste output. However, it is most often caused by the use of lead in plumbing systems.

Lead plumbing was almost universal some decades ago but has now largely been replaced by other metals and by plastic. Older properties still frequently have such systems even though, in the UK at least, local authorities are replacing them wherever possible.

The use of lead plumbing is an especially serious problem where the water supply is 'soft' with a low pH. Such water is plumbo-solvent and can dissolve large amounts of lead from the system. Hard water, on the other hand, of high pH and containing dissolved salts of calcium and magnesium, forms 'scale' within the system which prevents solution of lead and other metals.

Acid moorland water, such as is common in rural Scotland and Wales, may dissolve as much as 25 mg/litre of lead — several thousand times the WHO recommended limit. The danger is increased when water has been standing in contact with metal for several hours and not simply running rapidly through lead pipes. The UK Department of the Environment survey referred to above, found that 9 per cent of all homes tested had more than 100 μg/litre of lead and another 20 per cent more than 50 μg/litre (the WHO maximum permitted level) in 'first-run' tap water. These percentages were reduced to 4 and 10 per cent in 'daytime' running water. In the light of such differences between 'first-run' and 'daytime' tap-water it is good policy, especially where the feeding of young children is concerned, to allow water to run out of the tap for some minutes before using it to prepare food or beverages. The lead-enriched overnight water will flow away down the drain and fresh, less polluted water, will reach the tap.

This precaution is not a trivial matter, as H. Egan of the Laboratory of the Government Chemist of the UK noted[56]. He has calculated that 'an untreated private supply with the maximum amount of lead allowed would, on the assumption that 10 per cent of lead is absorbed and that two early-morning cups of tea are consumed, give a daily uptake of about 20 μg'.

In the modern home, lead pipes and tanks are seldom used and have been replaced by plastics. Though lead salts are used as stabilisers in the making of these polymer materials, there is no evidence that the metal can be leached out in dangerous amounts in normal use. However, in newly installed PVC systems, lead may appear when water first runs through. It is advisable for safety that plastic plumbing systems should be thoroughly flushed out over a period of several hours before the water is used for human consumption.

Alcoholic Beverages

The use of lead pipes and other equipment in breweries and cider factories in the past was often responsible for contamination of beer and cider. Today's plants use stainless steel or other lead-free material, and significant levels of lead contamination are seldom encountered in commercially prepared alcoholic beverages. A study of a range of UK and imported bottled and canned lagers[57] found less than 0.06 mg/litre of lead in nearly 150 samples examined. Nevertheless, as we have seen above, a substantial proportion of total lead intake in Australia and Germany results from beer drinking. Even if the concentration of the metal in the beer is low, a high consumption of the beverage, as is common among males in both countries, leads to a considerable overall intake of the metal[58].

As was noted in an earlier chapter, beverages preserved in cans may take up lead from solder if it is used on lids or seams. However, most beverage cans, for beers and other alcoholic products, are of extruded aluminium and consequently lead contamination is not a problem. Even where tinplate and lead solder is used, modern production techniques are such that lead contamination of the can contents seldom occurs.

Contamination of wines with lead, especially if they are prepared in traditional ways, is not uncommon. The lead pick-up can occur in a number of ways. Most commonly it results from the use of lead foil caps or covers over the cork and bottle top. A move to replace lead foil with plastic has reduced this danger, but reports of lead in wines from this cause are still seen in the literature[58]. The use of lead and lead alloys in construction and repairing of fermentation and other equipment has also been known to result in contamination.

Several recent reports have commented on the relatively high-level contribution that can be made by wine drinking to lead intake. Lead in wines sold in Italy, West Germany and the UK were all found to be generally in the order of 50–100 μg/litre[59], which is considerably higher than that normally found in other beverages. A correlation has been demonstrated between wine consumption and blood-lead levels[60].

Contamination of potable spirits on a smaller scale due to the use of tinned measures has been reported on a number of occasions. One such measure tested by the author was found to release 5 mg/litre of lead when filled with a dilute acetic solution for less than one hour. Lead-crystal decanters have been called a health risk in a recent letter to the Editor of the *Australian Medical Journal*[61]. The writer found that wine, including port and sherry, in such decanters, gained in lead concentration with time.

Concern about lead contamination of alcoholic beverages is not new. In 1767, Sir George Baker, in his *Essay on the Endemic Colic of Devonshire*, blamed the use of lead-lined troughs in the manufacture of cider, and the resulting lead contamination of the beverage, for the illness. Almost 200 years later, in 1954, the British Food Standards Committee reported that lead was one of the most widespread and serious of metallic contaminants of drink. This fact the Committee attributed largely to the use of lead and lead alloys in the equipment used in the industry, as well as to the use of lead insecticides in agriculture. In 1961 a general limit of 1.0 mg/litre of lead in beer, cider, and sherry was introduced. Later regulations reduced this permitted level. The Committee also recommended that the use of lead piping and lead-containing equipment should be stopped. As we have seen, these and similar regulations in other countries, have not yet entirely solved the problem.

Lead contamination of beverages still occurs frequently today in home production. Because of the use of readily available, but not necessarily food-quality equipment, illicit spirits are often found to contain large amounts of lead and other toxic metals. Poisoning from the consumption of such beverages is not uncommon[62]. The presence of lead in a product is usually related to the use of lead pipes or solder in the apparatus.

Lead poisoning may also result from storage of alcoholic and other fluids in unsuitable metal or glazed earthenware containers. Fatalities due to this latter cause have been reported in recent years. Apple juice stored in an earthenware jar caused the death by lead poisoning of a young boy in Canada[63]. In 1960 a case of lead poisoning due to drinking home-made wine which had been stored in an earthenware container was reported in Britain[64]. Forty persons were poisoned in a similar manner in Jugoslavia[65]. An American medical doctor suffered lead poisoning from the constant use of a ceramic mug made by his son in a craft class[66].

In spite of numerous warnings by health authorities of the dangers of using poorly glazed vessels as food and drink containers, and the regulations that have been introduced requiring high standards of manufacture, lead and other heavy metal poisonings resulting from their misuse continue to be reported. A case of lead encephalopathy in a 33-year-old woman who regularly drank cider from an Italian-made Toby mug, was reported recently from the UK[67]. A study in Spain showed that the use of vitrified and non-vitrified earthenware to hold wines, vinegars and traditional pickling liquids continued, with, as a result, high blood levels of lead and poisonings among the consumers of the products[68]. Clearly, the comment made some 20 years ago by Klein and his colleagues in their report on lead poisonings from earthenware containers in Canada is still valid:

Though the reported frequency of lead poisoning from pottery has been low, the true figure may be considerably higher. Fortunately, relatively few earthenware containers are used for storage of acidic solutions, and many are used only for decorative purposes. Public demand for hand-made pottery is leading to increased production and availability. Unless this is matched by an increased awareness of the problem by potters and governments, one can expect to see an increase in lead poisoning from this source[63].

Contamination of foods and beverages due to the use of utensils made from unsuitable metals has already been discussed in an earlier chapter. Tin, which contains lead, and is used to surface copper and iron utensils, as well as ceramic and other glazes on cast-iron casseroles, may, as we have seen, cause contamination under certain conditions. Pewter, used particularly for tankards and other vessels for alcoholic beverages, can be a particular problem. An Australian manufacturer of pewter and pewter-lined jugs and other vessels, sold primarily as souvenirs to tourists, was obliged to recall his products, following a case of lead poisoning when one of the jugs was used to hold orange and tomato juices. The inner lining of the jug was found to contain a 60–40 lead–tin mix[69].

ANALYSIS OF FOODSTUFFS FOR LEAD

Both wet digestion with nitric acid and dry ashing can be used for sample pre-treatment for the estimation of lead in food. Use of sulphuric acid can result in low recoveries, due to the formation of insoluble sulphate. Perchloric acid causes loss in the ashing stage and is not suitable for electrothermal AAS. Recovery in dry ashing is good, provided the temperature is maintained below 500° C.

Atomic absorption spectrophotometry is the current method of choice for lead in foods. Since lead levels in the majority of biological samples are relatively low, and sensitivity of flame AA is poor, it is advisable to use electrothermal atomisation. Contamination is a problem because of the widespread distribution of lead, and strict laboratory hygiene must be practised. Matrix effects occur but can be overcome by preliminary separation of the lead from the digest or ash. This can be done by extraction into 4-methylpentane-2-one (MIBK) using ammonium pyrrolidine dithiocarbamate (APDC) as a cleaning agent. The use of background correction also improves the technique. Other metal ions do not interfere with the estimation in an air-acetylene flame. High concentrations of anions, however, such as phosphate, acetate and carbonate, suppress lead absorbance significantly.

Details of the determination of levels of lead in canned foods using flameless AAS are given in a recent Swedish study[70]. The foods were ashed at 450° C and the ash dissolved in 1M HCl. The solution was diluted with 0.1M HNO_3 to reduce matrix interference. Graphite furnace AAS, with a deuterium lamp for background correction, was used for analyses. It was necessary, because of very severe matrix interference met with in the determination of lead in canned beans and peas, to extract the solutions with APDC/MIBK after adjusting the pH to between 1 and 2 with NaOH, followed by back extraction into HNO_3.

Lead was determined directly in wine and other alcoholic beverages, with no pretreatment apart from acidification, by Sherlock and his colleagues, using graphite furnace AAS[58]. Triton X and ammonium dihydrogen phosphate were used for matrix control.

REFERENCES

1. AGRICOLA, G. (1556). *De Re Metallica*, English translation by H.C. and L.H. HOOVER (Dover, New York, 1950).
2. GRANDJEAN, P. (1978). Widening perspectives of lead toxicity. A review of health effects of lead exposure in adults. *Environ. Res.*, 17, 303–21.
3. AUSTRALIAN LEAD DEVELOPMENT ASSOCIATION (1988). Metals review. *Elements*, 16, 7.
4. DOWDING, M.F. (1978). Lead. *Met. Materials (UK)*, July, 27–36.
5. PERRIN, J. (1974). Le plom du chasse. *Materiaux et Techniques*, November, 518–20.
6. RATHUS, E.M. (1958). The history of lead in Queensland. *Ann. Rep. Qld. State Dept. Health*, 114–18.
7. POLLOCK, C.A. and IBELS, L.S. (1988). Lead intoxication in Sydney Harbour bridge workers. *Aust. NZ. J. Med.*, 18, 46–52.
8. AUSTRALIAN ACADEMY OF SCIENCE (1981). *Health and Environmental Lead in Australia* (Canberra).
9. SCHROEDER, H.A. and NASON, A.P. (1971). Trace element analysis in clinical medicine, *Clin. Chem.*, 17, 461–74.
10. BARRY, P.S. (1975). Lead concentration in human tissues. *Br. J. Indust. Med.*, 32, 119–39.
11. KEHOE, R.A. (1961). The metabolism of lead in man in health and disease. *J. Ropy. Inst. Pub. Health Hyg.*, 24, 101–20; 129–43; 177–203.
12. ALEXANDER, F.W., DELVES, H.T. and CLAYTON, B.E. (1973). *Environmental Health Aspects of Lead*, 319–31 (Commission of European Communities).

13. HOTIUCHI, K.(1970). Lead in human tissues. *Osaka City Med. J.*, 16, 1–28.
14. DABEKA, R.W. and MCKENZIE, A.D. (1988). Lead and cadmium levels in commercial infants foods and dietary intake by infants 0–1 year old. *Food Add. Contam.*, 5, 333–42.
15. SCHROEDER, H.A. and TIPTON, I.H. (1968). Lead in ancient and modern skeletons. *Arch. Environ. Health.*, 17, 965–78.
16. WEISS, D., WHITTEN, B. and LEDDY, D. (1972). Lead levels in hair of American children. *Science*, 178, 69–70.
17. PETERING, H.G., YEAGER, D.W. and WITHERUP, S.O. (1973). Hair analysis. *Arch. Environ. Health*, 27, 327–30.
18. REILLY, C. and HARRISON, F. (1979). Zinc, copper, iron and lead in scalp hair of students and non-students in Oxford. *J. Hum. Nutr.*, 33, 250–4.
19. NATIONAL ACADEMY OF SCIENCE (1972). *Airborne Lead in Perspective* (National Research Council, Washington, DC).
20. REILLY, C. and REILLY, A. (1972). Patterns of lead pollution in the Zambian environment. *Med. J. Zambia*, 6, 125–7.
21. WORLD HEALTH ORGANISATION (1976). *Environmental Health Criteria* 3: *Lead.* (WHO, Geneva).
22. INGLIS, P., HENDERSON, D.A. and EMMERSON, B.T. (1978). Lead-related nephritis. *J. Pathol.*, 124, 65.
23. COMMISSION OF THE EUROPEAN COMMUNITIES (1977). Council Directive of 29 March on biological screening of the population for lead. *Off. J. EEC*, 20, L105/10–105/17.
24. QUINN, M.J. (1985). Factors affecting blood lead concentrations in the UK: results of the EEC blood lead surveys, 1979–1981. *Int. J. Epidem.*, 14, 420–31; ENGLERT, N., KRAUSE, C.H., THRON, H.L. and WAGNER, M. (1987). Studies on lead exposure of selected population groups in Berlin. *Trace Elem. Med.*, 4, 112–16.
25. EASTWELL, H.D. (1979). Petrol-inhalation in aboriginal towns. *Med. J. Aust.*, 2, 221–4.
26. COULEHAN, J.L., HIRSCH, W., BRILLMAN, J. *et al.* (1983). Gasoline sniffing and lead toxicity in Navajo adolescents. *Pediatrics*, 71, 113–17.
27. BROWN, A. (1983). Petrol sniffing lead encephalopathy. *New Zealand Med. J.*, June, 421–2.
28. US DEPARTMENT OF HEALTH, EDUCATION AND WELFARE (1965). *Survey of lead in the atmosphere of three urban communities, P. H. S. Pub.*, 999 – AP – 12 (Cincinnati Public Health Service).
29. BEAUD, P., ROLLIER, H. and RAMUZ, A. (1982). Contamination by road traffic of foods sold from shop-fronts. *Mitteilunaen aus dem Gebiete der Lebensmitteluntersuchung und Hygiene*, 73, 196–207.
30. TJELL, J.C., HORMAND, M.F. and MOSBAEK, H. (1979). Atmospheric lead pollution of grass grown in a background area in Denmark. *Nature (London)*, 280, 425–6.
31. WYLIE, P.B. and BELL, L.C. (1973). The effect of automobile emissions on the lead content of soils and plants in the Brisbane area. *Search*, 4, 161–2.
32. WILLIAMS, C. (1974). The accumulation of lead in soils and herbage at Rothamsted Experimental Station. *J. Agric. Sci.. Camb.*, 82, 189–92.
33. FOOD AND AGRICULTURAL ORGANISATION/WORLD HEALTH

ORGANISATION (1972). Sixteenth Report of the Joint FAO/WHO Expert Committee on Food Additives. *WHO Tech. Rep. Ser. No.* 505 (WHO, Geneva).

34. LOUEKARI, K. and SALMINEN, S. (1986). Intake of heavy metals from foods in Finland, West Germany and Japan. *Food Add. Contam.*, 3, 355–62.

35. KEHOE, R.A. (1961). The metabolism of lead in man in health and disease. *J. Roy. Inst. Pub. Health Hyg.*, 24, 1–96, 101–20, 129–43, 179–203.

36. DEPARTMENT OF HEALTH AND SOCIAL SECURITY (1980). *Report of a DHSS Working Party on Lead in the Environment* (HMSO, London).

37. WORLD HEALTH ORGANISATION (1976). *Environmental Health Criteria.*, 3, *Lead* (WHO, Geneva).

38. HORIUCHI, K. (1970). Lead in the environment and its effects on men in Japan. *Osaka City Med. J.*, 16, 1–28.

39. HORIGUCHI, S., TERAMOTO, K., KURONO, T. and NINOMIYA, K. (1978). An attempt at comparative estimate of daily intake of several metals (As, Cu, Pb, Mn, Zn) from foods in thirty countries in the world. *Osaka City Med. J.*, 24, 237–42.

40. AUSTRALIAN ACADEMY OF SCIENCE (1981). *Health and the Environment in Australia*. (Australian Academy of Science, Canberra).

41. REILLY, C. (1985). The dietary significance of iron, zinc, copper and lead in domestically prepared food. *Food Add. Contam.*, 2, 209–15.

42. ALEXANDER, F.W., DELVES, H.T. and CLAYTON, B.E. (1973). The uptake and excretion by children of lead and other contaminants, in *Proc. Intern. Symp. Environ. Health Aspects of Lead*, Amsterdam 1972, Luxembourg, 219–31. Commission of the European Communities, 1973.

43. NATIONAL ACADEMY OF SCIENCES (1972). *Airborne Lead in Perspective* (National Research Council, Washington, DC).

44. RICE, D.C. (1986). Behavioural toxicity of lead in monkeys. *Food Chemical Codex Workshop on Lead*. Nat. Res. Council Commission on Life Sciences, Mont Ste. Marie, Quebec, August, 14–15.

45. MAHAFFEY, K.R. (1983). Absorption of lead by infants and young children, in *Health Evaluation of Heavy Metals in Infant Formula and Junior Food*, eds SCHMIDT, E.H.F. and HILDEBRANDT, A.G., 69–95 (Springer Verlag, Berlin).

46. NUTRITION FOUNDATION'S EXPERT ADVISORY COMMITTEE (1982). *Assessment of the Safety of Lead and Lead Salts in Food* (Nutrition Foundation, Washington, DC).

47. DABEKAR, W. and MCKENZIE, A.D. (1988). Lead and cadmium in commercial infants food and dietary intake by infants 0–1 year old. *Food Add. Contam.*, 5, 333–42.

48. SHERLOCK, J.C. and QUINN, M.J. (1986). Relationship between blood lead concentrations and dietary lead intake in infants: the Glasgow Duplicate Diet Study 1979–1980. *Food Add. Contam.*, 3, 167–76.

49. EUROPEAN ECONOMIC COMMUNITY (1980). Council Directive of 15 July 1980 relating to the quality of water intended for human consumption (80/778/EEC). *Off. J. European Communities*, L239/11–29, 31 August 1980.

50. DEPARTMENT OF THE ENVIRONMENT (1977). *Lead in Drinking Water. A Survey in Great Britain* 1975–1976. Pollution Paper No. 12 (HMSO, London).

51. VOS, G., HOVENS, J.P.C. and DELFT, W.V. (1987). Arsenic, cadmium, lead and mercury in meat, livers and kidneys of cattle slaughtered in the Netherlands during 1980–1985. *Food Add. Contam.*, 4, 73–88.

52. MINISTRY OF AGRICULTURE, FISHERIES AND FOOD (1975). *Food Additives and Contaminants Committee Review of the Lead in Food Regulations.* FAC/REP/21 (HMSO, London).

53. NATIONAL RESEARCH COUNCIL OF CANADA (1973). *Lead in the Canadian Environment.* NRCC No. 13682 (Ottawa).

54. SREBOCAN, V., POMPE-GOTAL, J., SREBOCAN, E. and BRMALJ, V. (1983). Heavy metal contamination of the ecosystem Crna Mlaka II. Lead concentrations in tissues of Mallard duck with and without ingested lead shot present in gizzards. *Veterinarski Arhiv.*, 53, 259–66.

55. CRAUN, G.F. and MCCABE, L.J. (1975). Lead in domestic water supplies. *J. Am. Water Workers Assoc.*, 67, 593–9.

56. EGAN, H. (1972). *Lead in the Environment*, ed. HEPPLE, P., 34–42 (Institute of Petroleum, London).

57. BINNS, F., ENZOR, R.J. and MACPHERSON, A.L. (1978). Lead contamination of beers. *J. Sci. Food Agric.*, 29, 71–4.

58. SHERLOCK, J.C., PICKFORD, C.J. and WHITE, G.F. (1986). Lead in alcoholic beverages. *Food Add. Contam.*, 3, 347–54.

59. JORHEM, L., MATTSSON, P. and SLORACH, S. (1988). Lead in table wines on the Swedish market. *Food Add. Contam.*, 5, 645–9.

60. ELINDER, C.G., LIND, B., NILSSON, B. and OSKARSSON, A. (1988). Wine — an important source of lead exposure. *Food Add. Contam.*, 5, 641–4.

61. DE LEACY, E.A. (1988). Lead-crystal decanters — a health risk? *Aust. Med. J.*, 147, 622.

62. OH, S.J. (1975). Heavy metals poisoning from illicit alcoholic beverages. *Arch. Phys. Med. Rehabil.*, 56, 312–7.

63. KLEIN, M., NAMER, R., HARPUR, E. and CORBIN, R. (1970). Earthenware containers as a source of fatal lead poisoning. *New Eng. J. Med.*, 283, 669–72.

64. WHITEHEAD, T.P. and PRIOR, A.P. (1960). Lead poisoning from earthenware container. *Lancet*, 1, 1343–4.

65. BERITIC, T. and STRAHULJAK, D. (1961). Lead poisoning from glazed surfaces. *Lancet*, 1, 669.

66. HARRIS, R.W. and ELSEA, W.R. (1967). *J. Am. Med. Assoc.*, 202, 544–6.

67. ZUCKERMAN, M.A., SAVORY, D. and RAYMAN, G. (1989). Lead encephalopathy from an imported Toby mug. *Postarad. Med. J.*, 65, 307–9.

68. MENDEZ, J.H., DE BLAS, O.J. and GONZALEZ, V. (1989). Correlation between lead content in human biological fluids and the use of vitrified earthenware containers for foods and beverages. *Food Chem.*, 31, 205–13.

69. MONTGOMERY, B. (1988). Poisonous jugs recalled. *The Australian*, 8 November.

70. JORHEM, L. and SLORACH, S. (1987). Lead, chromium, tin, iron and cadmium in foods in welded cans. *Food Add. Contam.*, 4, 309–16.

CHAPTER 6

MERCURY AND CADMIUM

MERCURY

Mercury is one of the ancient metals. It is relatively easily extracted from its ores and, in spite of its limited practical uses, has been sought and valued for many hundreds of years. The Spanish conquistadors brought it back to Europe from the New World to augment supplies from their own rich mine at Almaden in Spain.

It is, in several respects, a peculiar metal, useless for making weapons or tools, for it is a liquid. The ancients called it 'quicksilver' and valued it for its medicinal properties — some real, some imagined. They also found that its physical properties could be changed by forming it into an amalgam with other metals. A most valuable use was for 'silvering' mirrors. For the medieval alchemist, mercury had its own peculiar value and it played an important part in the search for the 'philosopher's stone' which would allow transmutation of base metals into gold.

Today, mercury is valued and widely used in industry, mainly because of its chemical properties. It is used as a catalyst in a variety of industrial and laboratory reactions, some of great economic value. Its physical property of high conductivity makes the liquid metal valuable in the electrical industry.

From early times men knew that mercury had sinister side-effects. Georgius Agricola wrote of the 'quicksilver disease' from which miners, who extracted mercury from its ores in the Hartz Mountains, suffered. Others who worked with the metal were also known to develop distinctive mercury illnesses. The makers of felt hats who used mercury to prepare the material, showed peculiar mental symptoms which gave rise to the saying 'mad as a hatter'. The term hatter, suffered a strange transportation from Lewis Carroll's 'Alice in Wonderland' to nineteenth-century Australia where gold miners, affected by the mercury they used to extract the metal from crushed ore, were known as 'hatters'. Those gold mines, now worked out and long abandoned, still cause environmental and, at least to aquatic life, health problems. In the Australian State of Victoria, the Lederderg River, on the banks of which gold was once extracted by the mercury amalgamation method, has mercury-contaminated silt and fish with elevated levels of the metal in their tissues[1].

Though hazards of this kind have been reduced almost everywhere by improved working conditions, we are still very much aware of the potential danger for health that the metal poses in many of its applications. Of particular significance in recent years have been the serious instances of poisoning, especially by organic mercury, through industrial pollution and misuse of certain compounds. Mercury can still be considered among the most dangerous of all the metals we are likely to meet in our food. Worldwide attention among food and toxicology experts is focused on the metal, and regular monitoring of levels in human diets is carried out in many countries. An important World Health Organisation document, which draws attention to the problem, has recently been published[2]. The document considers such topics as persistence of mercury in the environment, bioaccumulation and transport in the aquatic chain, levels in a variety of foods and implications for human mercury consumption.

CHEMICAL AND PHYSICAL PROPERTIES

Mercury, given the symbol Hg after its Greek name *hydrargyrum* (liquid or 'quick' silver), is element number 80 in the Periodic Table. Its atomic weight is 200.6 and its density 13.6, making it one of the heavier of the metals. It is liquid over a wide range of temperatures, from its melting point of $-38.9°$ C to its boiling point of $356.6°$ C. Its oxidation states are 1 and 2.

Elemental mercury is rather volatile. A saturated atmosphere of the vapour contains approximately 18 mg/m^3 at 24° C. The metal is slightly soluble in water and lipids. Its solubility in the latter is of the order of 5–50 mg/litre. Sulphates, halides and nitrates of mercury are also soluble in water. An equilibrium is formed between the Hg^0, Hg^{2+} (mercuric) and Hg_2^{2+} (mercurous) states in aqueous solution. The proportions of the different oxidation states are determined by the redox potential of the solution and the presence of complexing substances. Mercury can form many stable complexes with biological compounds, especially through —SH (sulphydryl) groups. In aqueous solution, four different combinations with chlorine are formed: $HgCl^+$, $HgCl_2$, $HgCl_3^-$ and $HgCl_4^{2-}$. Mercurous mercury is rather unstable and tends to disassociate in the presence of biological materials to give one atom of metallic mercury and an Hg^{2+} ion.

ORGANIC COMPOUNDS OF MERCURY

Volatile compounds are formed between alkyl mercuric compounds and the halogens. These are highly toxic. Less volatile are the hydroxide and nitrate of the short-chain alkyl mercuric compounds. Methyl and ethyl mercury chloride have a high solubility in solvents and lipids. The methyl mercury group has an affinity for sulphydryl groups. This is the linkage through which it can bind to protein in living organisms.

ENVIRONMENTAL DISTRIBUTION

Mercury is not very widely distributed in the environment. It is only about the sixtieth most abundant element in the earth's crust, with an average concentration of about 0.5 mg/kg and only a relatively small number of areas of concentration. The world's principal mercury mines occur in what is known as the 'mercuriferous belt' which stretches along the Mid-Atlantic Ridge, the Mediterranean, South and East Asia and the Pacific Ring, and is associated with volcanic activity.

Famous and ancient mines of Europe and the Western world are Idria in Jugoslavia, Almaden in Spain and New Almaden in California. There are also large deposits in the USSR. Apart from its extraction by man in mining operations, large quantities of mercury are released into the environment by volcanoes. This release can contaminate plants and animals and introduce mercury into the food chain[3]. Man's industrial activities, including the combustion of fossil fuels, also contribute to the environmental mercury burden.

Mercury occurs in the earth's crust mainly in the form of its sulphides. Cinnabar, the red sulphide of mercury, is the principal ore mined. It can contain upwards of 70 per cent of the metal and is relatively easily extracted by roasting in air. The mercury volatilises and is condensed and collected into flasks.

PRODUCTION AND USES

World production of mercury is about 10 000 tonnes each year. A quarter of this is used in electrodes in the chlor-alkali industry, another quarter in the electrical industry and the remainder in the manufacture of paints, a variety of instruments, agrochemicals and other specialist items. These include catalysts used, for example, in the manufacture of plastics, as 'slimecides' in the paper and pulp industry, and as germicides and pesticides. Until recently about 3 per cent of total production was used as mercury amalgam in dentistry. However, the introduction of newer materials as well as concern about the long-term effects on health of mercury tooth fillings, have reduced this use of the metal. Formerly, extensive use was also made of mercury, in organic form, as an antifungal dressing for seeds, but this use has decreased considerably following several serious poisoning incidents which will be discussed below.

In addition to the mercury produced by mining activities, it is estimated that a further 10000 tonnes are released into the environment by the combustion of coal, oil and gas, as well as by waste disposal and other industrial activities[4]. However, such production of the metal is far surpassed by natural production of upwards of 150000 tonnes in degassing from the earth's crust and oceans.

MERCURY IN THE ENVIRONMENT: CHEMICAL MODIFICATIONS

Metallic mercury undergoes a number of chemical changes when it is released into the environment. In soil, sulphur-reducing bacteria can convert it into its sulphide. The same transformation can occur under anaerobic conditions in water, but generally, in an aquatic system, aerobic methylation is the more important transformation. Methyl mercury, CH_3Hg^+, is formed by microorganisms from elemental as well as mercuric mercury. From the point of view of the microorganism, methylation is an efficient way of detoxifying inorganic mercury.

Methylation of mercury usually takes place in the upper layers of sediment on sea or lake bottoms. Methyl mercury so formed is rapidly taken up by living organisms. In this way it enters the food chain via plankton-feeders, and up through predators feeding on these. At the end of the chain we find the larger carnivorous fish, the marine sharks and freshwater pike, for instance, who, with growth in size and age, can accumulate dangerously high levels of organic mercury. After these come the human consumers.

As has been shown by the Swedish Expert Group[5], local loading of methyl mercury in the environment and organisms, can be due to industrial pollution, for example from paper pulp and chloralkali manufacture, or, as is found in the coastal waters between Australia and New Zealand, naturally-occurring mercuriferous deposits[6]. While little can be done, from the point of view of human health, with regard to mercury from natural sources, except to restrict the use of fish from the contaminated area, strict pollution control can be effective where industry is the source of contamination. It should be mentioned that another result of industrial pollution, acid rain, has been reported to increase the biological availability of methyl mercury in fish[7].

BIOLOGICAL EFFECTS OF MERCURY

Mercury is a cumulative poison which is stored mainly in the liver and kidney. The level of accumulation depends on the type of organism and the chemical form in which the mercury occurs. Inorganic mercury, in its metallic form, is unlikely to cause poisoning. The metal is poorly absorbed and what is taken in is quickly eliminated from the body. Mercury vapour, however, is readily absorbed from the lungs and is a much more serious problem than the liquid metal. It is capable of causing both acute and chronic poisoning.

Compounds of inorganic mercury can be much more toxic than the metal itself. Mercuric chloride, as its common name 'corrosive sublimate' indicates, is a highly toxic and dangerous substance. It has often been used as a poison in homicides and suicides. Though less than 10 per cent of the amount ingested is absorbed initially, this is capable of causing intestinal damage which can result in further absorption. The greater part of the absorbed Hg^{2+} is concentrated in the organs, especially the kidneys. It is a very painful and usually fatal form of acute poisoning, with symptoms of vomiting and kidney failure. Chronic poisoning is unlikely with Hg^{2+}, unless mercury vapour is also present.

The biological effects of organic mercury compounds are much more severe than those of either metallic or mercuric mercury. Of the different organic mercury compounds which occur naturally, methyl mercury and ethyl mercury are the most toxic. Methyl mercury has, in fact, been listed as one of the six most dangerous chemicals in the environment by the

International Program for Chemical Safety[8]. Both these short-chain organic compounds of the element have similar properties and behaviour. Methyl mercury ingested in food is efficiently absorbed in the intestine. It rapidly enters the bloodstream where it is bound to plasma protein. Most of the methyl mercury is initially accumulated in red blood cells and then is distributed to other tissues. The brain has a special affinity for organic mercury compounds and can accumulate methyl mercury to about six times its level in other tissues. More than 95 per cent of the mercury in brain has been shown to be in organic form. Methyl mercury is a well-established neurotoxin for both adults and the foetus[9].

Accumulation of methyl mercury in the foetus is similar to what occurs in the mother, though foetal brain levels of mercury may be higher. Prenatal poisoning has been shown to occur, even when the mothers of affected children show no clinical signs themselves of mercury toxicity. Methyl mercury appears in breast milk of lactating women who have ingested the substance.

Some of the absorbed methyl mercury is excreted from the body via the kidneys, but the bulk goes out in bile in the faeces. The biological half-time, that is the time it takes for half of an absorbed load to be excreted from the body, seems to be about 70 days, but this probably varies between individuals.

Once a toxic dose of methyl mercury has been absorbed it will be retained for a long time, causing functional disturbances and damage. There seems to be a latent period between absorption of the dose and appearance of symptoms. Clinical signs of methyl mercury poisoning are sensory disturbances in the limbs, the tongue and around the lips. With increasing intake the symptoms become more severe. The central nervous system is damaged irreversibly, resulting in ataxia, tremor, slurred speech, tunnel vision, blindness, loss of hearing and, finally, death.

Levels of mercury in blood can give a reliable index of the total methyl mercury burden of the body. Methyl and total mercury concentrations in hair are also used with success to measure levels of intake. Hair mercury measurements, in sections cut at different levels from the scalp, can provide a chronological record of blood mercury levels at different stages during a prolonged period of mercury poisoning[10].

An interesting aspect of mercury toxicity is that the element selenium appears to counteract both inorganic and organic mercury poisoning. This has been shown in the case of several animal species[11]. This function of selenium will be discussed in a later chapter. Here it will be sufficient to notice that selenium levels in vegetables grown on sludge-treated soil tend to be higher than normal. It is on such soils that plants often accumulate high levels of mercury and other toxic elements[12].

MERCURY IN FOOD

Mercury is found in food in three different forms: elemental mercury, mercuric mercury and alkyl mercury. The form influences absorption, distribution in body tissues and biological half-life. Normally the level of mercury, of any kind, in our diet, is very low. Concentrations of total mercury range from a few micrograms to 50 μg/kg[13].

In the absence of gross contamination of soil or irrigation water, the following mean levels of concentration can be expected in different food groups:

meat 10; fish 200; dairy products 5; vegetables 10; cereals 10; other foods 5 μg/kg (wet weight)[14].

Fish and organ meat, especially kidney and liver, have on occasion been found to contain much higher levels of mercury than those given here. This has been attributed to industrial pollution of waters or soil by, for example, use of sewage sludge as a soil improver. A range of 1.0–136 μg/kg of mercury was found in kidneys of Dutch cattle and 7.0–14.0 μg/kg in their livers[15]. Fish from the highly polluted Mediterranean coast of Egypt had total mercury contents ranging from 0.06 to 1.61 mg/kg[16]. The upper range is well above the maximum permitted levels for mercury in foods recommended by the Codex Alimentarius and adopted by many countries. Australian regulations, for instance, stipulate a top of 0.5 mg/kg in fish (mean level), and 0.03 mg/kg in all other foods[17].

Even vegetables can be affected by mercury contamination of the environment and accumulate unusually high levels of the element[18]. Lettuce grown on sludge-treated soil, for instance, was found to have more than 40 μg/kg (dry weight) of the element. Lettuce grown on uncontaminated soil would have ten times less mercury. While, in fact, the total amount of mercury accumulated in the leaves of the lettuce is well below the legal level of 30 μg/kg permitted, for example, in Australia, and would be unlikely to cause any health problem, it is significant that more than 16 per cent of that mercury was found to be in the form of methyl mercury.

The presence of organic mercury in food, as has been mentioned, can be a serious problem. This occurs, in many cases, when alkyl mercury is formed from the inorganic element in marine or freshwater sediments, and subsequently enters the food chain via filter-feeding organisms and their predators. This may lead to concentration by a factor of as much as 3000 from the water to the tissue of fish such as freshwater pike. Mercury is abundant in many polluted waters, which may contain as much as 200 ng/litre, compared to ocean water with 30 ng/litre[19].

Accumulation of mercury in fish is related to age and size. Large tuna over 60 kg in weight may have levels of organic mercury up to 1 mg/kg in muscle. This compares with terrestrial animals whose muscles rarely have more than 50 μg/kg and generally average 20 μg/kg. Fish living in highly polluted water may contain more than 10 mg/kg. In British coastal waters average levels of 210 μg/kg have been found, with, in the mouth of the Thames and in some stretches on the Lancashire coast, some levels up to 500–600 μg/kg[20]. Considerably higher levels of up to 6.9 mg/kg total mercury, of which 6.2 was in organic form, have been reported in South American tuna fish[21].

The notorious Minamata Bay tragedy in post-World War II Japan, was caused by mercury pollution of fishing grounds. What came to be known as 'Minamata disease' was officially recognised in 1956, but had actually been affecting local residents of the area for many years before that. The debilitating, and in several cases, fatal illness was found to be due to the consumption of fish contaminated by mercury discharged by a chemical factory on the shore of the bay. Levels of up to 29 mg/kg of methyl mercury were found in fish and shellfish. Though the cause of the disease, and the source of the pollution, were recognised in the mid 1950s the manufacturer continued to discharge mercury-contaminated effluent until 1968, and the tragedy continued[22]. The Bay remains polluted by organic mercury to this day, and levels in its fish are still high. By 1971 the total number of poisoning cases recorded in the area was 121, with 46 fatalities. Twenty-two of the cases were congenital. These were infants with cerebral involvement (palsy and retardation) born to mothers who had ingested contaminated fish.

The Japanese tragedy, which was followed by others also caused by industrial pollution elsewhere in Japan, caused considerable concern in other countries also where fish was consumed in quantity and mercury pollution occurred. In Sweden, though fish intake is generally about 30 g each day, some communities consume as much as half a kilogram daily. It was estimated that if the fish came from mercury-polluted waters, such an intake could result in an uptake of 5 mg of methyl mercury each day. Several well-fished lakes in Sweden, and in Canada, on which mercury-using pulp factories were situated, were known to be heavily polluted with the metal. Levels of methyl mercury of up to 10 mg/kg were found in the flesh of certain fish[23]. As a result, fishing was banned in some lakes in both countries and efforts were made to clear up and reduce the industrial pollution[24].

Though fish is the major source of methyl mercury in the diet, and there is evidence of a correlation between fish consumption and total dietary mercury intake[25], mercury intoxication has also been known to result from other causes. Alkyl mercury compounds were formerly widely used as antifungal agents for treating cereal seeds. Misuse of treated seeds, as food for human and animal consumption, resulted in a number of poisoning accidents, some of them on a very large scale. In Iraq, in 1960, seed wheat which had been treated with a methyl mercury fungicide and was donated by the US as assistance to allow planting to begin after a devastating drought had depleted seed stocks, was used to make flour. For several months bread made from the treated grain was eaten by thousands of people before symptoms of mercury poisoning began to appear. Many died, others were permanently incapacitated and there were many cases, also, of prenatal poisoning[26]. Several similar incidents, also caused by misuse of seed wheat treated with a mercurial fungicide, occurred in Pakistan and Guatemala. Even the use of mercurial-treated grain to feed pigs which were later butchered for pork has caused human poisoning[27]. Organic mercury seed dressings are now no longer used in the majority of countries and have been replaced by less toxic and persistent fungicides for such purposes.

ANALYSIS OF FOODSTUFFS FOR MERCURY

Two principal methods are used for the determination of both inorganic and organic mercury in foods: flameless, or cold vapour, atomic absorption spectrophotometry and gas chromatography. Where facilities are available, neutron activation analysis is also used, particularly for quality control purposes.

Flameless Atomic Absorption
A variety of analyser kit accessories are available to allow standard atomic absorption spectrophotometers to be adapted for cold vapour techniques. Both circulating and non-circulating methods are used. In the former, the mercury content of the sample is determined by measuring the transient absorbance produced when the mercury vapour is led through an absorbance cell. The cell replaces the usual burner in the light path. The

circulating method allows the progressive build-up of mercury until a constant absorbance is attained. Magos[28] has described a cold vapour AA method for the determination of both inorganic and organic mercury in biological samples. In this method mercury vapour is released from the samples, either after reduction in an alkali solution by $SnCl_2$ for inorganic mercury analysis or $SnCl_2/CdCl_2$ for total mercury analysis.

Gas Chromatography

A method has been described by Capon and Smith[29]. It involves an initial extraction of the samples with benzene, followed by several steps including acidification and treatment with ethanolic $Na_2S_2O_3$, with separation of the organic mercury into the benzene, leaving the inorganic mercury in the aqueous layer. The inorganic element is then converted into methyl mercury by treatment with methanolic tetramethyl tin. These steps are followed by gas chromatographic analysis. The technique requires considerable care and recovery at each stage must be monitored.

Neutron Activation Analysis

A method for the determination of total mercury in biological samples by radiochemical neutron activation analysis is described by Sjostrand[30]. The method has a high specificity and reliability. Neutron activation has the advantage of minimal sample preparation and avoids the errors which can easily occur of loss through volatilisation of mercury. It has been used effectively on small samples in total diet studies[31].

CADMIUM

Unlike mercury and lead, cadmium is not an 'ancient' metal, at least in terms of its use by man. It is only in recent years that it has found widespread industrial application, mainly in the metal plating and chemical industries. However, it is quite likely that it was unsuspectedly used, and that its highly toxic effects were experienced, by humans for many centuries. Cadmium is found associated with several other metals in ores and often contaminates the extracted metals. Its presence in zinc, for instance, is well known and in many instances in which 'zinc poisoning' was believed to have occurred, it was probably cadmium that was the toxic agent. Even minute amounts of cadmium are sufficient to cause poisoning. Moreover, since the metal is soluble in organic acids, it easily enters acid foods with which it comes in contact.

Cadmium is a highly toxic element. It has been described as 'one of the most dangerous trace elements in the food and environment of man'[32]. The extreme danger that can result from cadmium ingestion was brought to the world's attention, with horrifying impact, by the *itai-itai* disease outbreaks in Japan in the late 1960s. This tragedy, which was due, as was the case with several of the lead and mercury poisonings described above, to industrial pollution, and resulted in considerable human suffering and death, stimulated a flurry of investigations and alerted health authorities to the serious nature of the problem of cadmium pollution. It is now widely accepted that the metal is among the most dangerous of all the metal contaminants of food, not only because of its high level of toxicity, but also because of its wide distribution and its many important industrial applications. As is the case with mercury and lead, cadmium levels in the food supply and in water, are monitored by health authorities and permitted levels are regulated by legislation in many countries.

CHEMICAL AND PHYSICAL PROPERTIES

Cadmium is element number 48 in the Periodic Table, with an atomic weight of 112.4. It is a fairly dense (specific gravity 8.6), silvery white, malleable metal which melts at 320.9° C and boils at 765° C. It has an oxidation state of 2. Cadmium forms a number of water-soluble, inorganic compounds, including the chloride, sulphate and acetate. The sulphide is, however, only slightly soluble, while the oxide is insoluble. A number of organocadmium compounds have been synthesised, but these are very unstable and none have been found to occur naturally. The element forms complexes with organic compounds, such as dithiozone and thiocarbamate, and this property is the basis of colorimetric determination methods. It also joins to protein molecules through sulphydryl links.

An important physical property of cadmium is its ability to absorb neutrons readily. This makes it very useful for the manufacture of control rods in nuclear reactors.

PRODUCTION AND USES

Though cadmium sulphide occurs naturally as a yellow ore known as 'greenockite', this is not commercially exploitable. The metal is obtained as a by-product in the extraction and refining of other metals with which it is naturally associated. Zinc is the principal metal with which cadmium occurs, but it is also found with copper and lead. World production is relatively small, at around 2000 tonnes each year, but is increasing as new industrial uses of the metal are found.

One of the major uses of cadmium is as an anti-rust coating on iron. In the US upwards of 60 per cent of all cadmium is used for electroplating[33]. Cadmium provides a better surface protection than does zinc plate, and consequently it is widely used in the automobile industry for protecting engine and other components. Formerly it was also used for plating food and beverage containers, especially where they might be exposed to damp conditions, as in cold rooms and refrigerators. However, a number of poisoning incidents from food stored in such containers alerted authorities to the toxicity of cadmium plating and the metal is no longer used for this purpose.

Cadmium compounds are used as stabilisers in plastics. Cadmium sulphide and cadmium sulphoselenide are important as pigments in paints and plastics.

Cadmium forms fusible alloys with a number of other metals and these are widely used industrially. Cadmium–copper is used in high-conductivity cable wires, in bearing alloys and in automobile components. The metal is also used in some solders. An important application is as an electrode component in alkaline accumulators.

CADMIUM IN FOOD AND BEVERAGES

While cadmium is found in most foodstuffs, this is normally at a very low level of concentration, unless contamination has occurred.

The results of an Australian 'market basket survey'[34] are typical of levels found in foods in most countries. The overall range of the element in food in daily family use was 0.095–0.987 mg/kg, with a mean of 0.469 mg/kg. No samples had more than 1.0 mg/kg. Levels in individual foods were: bread <0.002–0.043; potatoes <0.002–0.051; cabbage <0.002–0.026; apples <0.002–0.019; poultry <0.002–0.069; minced beef <0.002–0.028; kidney (sheep) 0.013–2.000; prawns 0.017–0.913 mg/kg. Similar levels were found in a recent Canadian study[35].

The Australian results indicate that meat offal and seafoods tend to be higher in cadmium than other foodstuffs. This is a pattern that has been observed in other studies. Cadmium levels in animal tissues, especially liver and kidney, are strongly related to levels in animal feeds[36]. Sewage sludge used as a soil improver, as has already been mentioned, is a particularly important source of accumulated cadmium in animal tissues[37], as are also certain fertilisers such as superphosphate[38].

Industrial and other forms of contamination can also result in higher than average concentrations of cadmium in other foods, including cereals. Rice, in polluted areas of Japan, where *itai-itai* disease occurred, had approximately 1 mg/kg of the metal, compared to levels of 0.05–0.07 mg/kg in non-polluted regions. This has been attributed to cadmium released from non-ferrous metal mines and smelters which has resulted in '9.5 per cent of paddy soils, 3.2 percent of upland soils and 7.5 per cent of orchard soils in Japan being severely contaminated with cadmium'[39].

In Australia cadmium accumulation in agricultural soils and pasture plants from the use of cadmium-containing fertiliser is causing alarm. Phosphate from the Pacific Island of Nauru, which is widely used, contains between 70 and 90 mg/kg of the metal. It is believed that this has resulted in high levels of cadmium in wheat as well as in beef and sheep liver and kidney[40]. A similar problem has been recognised in Sweden[41]. That soil contamination can also result in build up of cadmium in other vegetable foodstuffs was shown in an earlier study by John[42].

Fish and other seafoods, from polluted waters, may contain high levels of cadmium. Recent Australian findings of an average of 0.42 mg/kg in crab meat, with a highest value of 2.3 mg/kg, have been reported[43]. Lisk, in the US, reported a range of 0.05–3.66 mg/kg in seafoods[44]. It is of significance that in Japan it has been found that 21.4 μg out of a total daily cadmium intake of 60 μg, was due to fish and seafood consumption[45].

A study of heavy metal intake by Canadian infants[46] found that cadmium intake was strongly influenced by whether or not soya-based formulas were used. Intake by infants consuming such formulas was more than three times that of others fed human or cows' milk. Actual intake by infants fed the soya-based formulas was approximately 0.5 μg/kg/day. Though this is well below the FAO/WHO PTDI for adults of 0.96–1.2 μg/kg, it is nevertheless significantly higher than intake by breast-fed infants and warrants serious consideration of the appropriateness of soya in infant formulas. As we will see, the same problem arises with aluminium in soya-based infant formulas.

Water normally makes a minimal contribution to dietary cadmium intake. Levels are usually less than 1 μg/litre. However, contamination can occur from the use of zinc-plated (galvanised) pipes and cisterns and result in an excessive intake of cadmium. A study of water used in boilers in several Scottish hospitals found that cadmium in some cases was significantly higher than the WHO standard of 10 μg/litre[47]. The maximum level recorded was 21 μg/1itre[48]. An Australian study found that while municipal water supplies were below the WHO limit, beverages made from this water in a continuous heating unit, in a university staff room, had significantly elevated cadmium levels[49]. In both cases the cadmium was attributed to the use of impure zinc in the construction of the water containers and heaters.

Since zinc normally contains some cadmium, it was suspected that the use of zinc-plated, or galvanised roofs and tanks to collect rainwater, as is the practice in many parts of the world, could expose consumers to cadmium toxicity. Another Australian study looked at this possibility[50]. Zinc used for galvanising iron in Australia is produced primarily by the electrolytic process. The cadmium content of zinc produced by this means has decreased markedly since the 1920s, from about 100 mg/kg to less than 10 mg/kg today. Cadmium levels in water held in tanks was found to correlate significantly with the cadmium content of the zinc used for the galvanising. No tank had levels above the WHO standard. However, the two oldest tanks examined had levels of 2.3 and 3.6 μg/litre, with all the others less than 1 μg/litre. It was concluded that there was little risk to the rural community in using tank water, even when this was run off galvanised roofs, provided the tanks were not more than 40 years old.

Other beverages may also have elevated levels of cadmium due to contact with cadmium plate. Soft drinks dispensed from vending machines with cadmium-plated parts, have been found to contain up to 16 mg/litre of the metal. Illicit alcoholic beverages, made in crude distillation apparatus, had up to 38 mg/litre[51].

Daily intake of dietary cadmium by adult Canadians has been estimated to be 13.8 μg (range 7.1–34 μg/day) or 0.21 μg/kg/day (range 0.10–0.49 μg/kg/day). These figures are not very dissimilar from those published for several countries which range from an overall average of 10–80 μg/day[52]. It is notable that the more recent Canadian figures are lower than those found in earlier Canadian and US studies, which reported 67 and 80 μg/day. It is probable that the main reason for the difference is not a reduction in pollution but improved analytical techniques.

The mean intake by the adult Canadians in the more recent study (0.21 μg/kg/day) is one-fifth of the FAO/WHO provisional tolerable daily intake of 0.96–1.2 μg/kg[53].

Several factors, apart from food contamination, may increase intake of cadmium. Cigarette smoking has been shown, for instance, to increase cadmium absorption by the body[54]. Cigarettes can have a cadmium content of between 0.9 and 2.0 μg/g (dry weight) and the element is readily absorbed through the lungs[55]. It has been estimated that for every 20 cigarettes smoked, 0.5–2 μg of cadmium can be absorbed in this way.

UPTAKE AND ACCUMULATION OF CADMIUM BY THE BODY

About 6 per cent of cadmium ingested in food and beverages is believed to be absorbed by the body under normal conditions. The presence of certain other substances in the diet, including calcium and protein, may increase the level of absorption. In the body, cadmium is bound mainly to metallothionein, a low molecular weight protein. This protein is involved both in the transport and in selective storage of a variety of metals and can bind up to 11 per cent of total metals in the body. Cadmium and zinc, and, to a lesser extent, iron, mercury and copper, compete for binding sites on metallothionein which is rich in sulphydryl groups. It is produced mainly in the liver, but also in the gastro-intestinal mucosa and the renal cortex in response to the presence of cadmium as well as of zinc. The dual affinity for the two metals probably accounts for the relationship between their metabolism which will be discussed below[56].

Most of the cadmium absorbed by the body is retained, with a little excreted through the kidneys or faeces. As a result of efficient retention, the biological half-life of cadmium is long, perhaps as much as 40 years. This retention has considerable health consequences. Even at low levels of exposure, the body accumulates the metal throughout life[57].

Newborn babies have very little cadmium in their tissues. It is well established that renal concentration steadily increases with age, reaching a maximum about the age of 50[58]. By this age, an American, East German or Swedish male can expect to have up to 30 mg of the metal in his body. The kidney will contain 25–50 mg/kg of this. A Japanese of the same age might be expected to carry a much greater load of the metal, with up to 100 mg/kg in kidney cortex. The level will be even higher if the Japanese also smokes[59]. It is at this maximum or 'critical' level of accumulation, about 50 years of age, that kidney damage is most likely to occur.

Methods for the biological monitoring of cadmium exposure are still subject to debate. Blood levels may reflect more recent exposure, but probably do not give an accurate indication of long-term kidney accumulation[60]. Concentrations in urine have been found, in certain cases, to reflect levels of accumulation in kidneys[27]. Levels of cadmium in hair are also used as indicators of exposure to the metal[61]. An interesting UK study of dietary cadmium intake and accumulation in the body, concludes that it is possible to predict the average cadmium concentration in the human kidney for a person aged 40–60 years from a knowledge of their dietary intake[62].

EFFECTS OF CADMIUM ON HEALTH

The ingestion of cadmium in food and drink can result rapidly in feelings of nausea, vomiting, abdominal cramp and headaches. Diarrhoea and shock can also occur. About 15 mg/litre of the metal in a liquid are enough to bring about these responses.

Long-term ingestion of cadmium causes serious renal damage, as well as bone disease leading to brittleness and even collapse of the skeleton[63]. These latter were among the most obvious and alarming of the symptoms seen in the notorious '*itai-itai*' disease outbreak in the Jintzu area of Japan. This occurred when cadmium-rich mine tailings were discharged into water used to irrigate paddy fields. Similar symptoms may also be observed in industrial workers chronically exposed to the metal.

Itai-itai disease is an osteomalacia and results from a combination of disturbances in calcium metabolism and tubular disfunction of the kidneys due to cadmium poisoning. It can be exacerbated, as occurred in *itai-itai* victims in Japan, by an inadequate calcium intake in the diet. These were generally poor peasants whose staple food was rice, with little meat or dairy products. Initially the illness manifested itself through pain in the back and legs. Pressure on bones, especially long bones in the legs and on the ribs, caused increasing pain. As the disease progressed, even slight bumps and knocks began to cause bone fractures. The name *itai-itai* expresses in a macabre fashion the sufferers' reaction to their condition, and means 'ouch-ouch'.

There is some evidence that cadmium exposure may result in a higher than normal level of cancer. This has been shown in the case of industrial workers exposed to vapour of cadmium oxide[64]. There is as yet no convincing evidence that ingestion of the metal will have a similar effect. Both genetic and teratogenic effects have been attributed to cadmium and chromosome aberrations have been observed in some *itai-itai* victims. Congenital abnormalities have occurred in rats fed cadmium-contaminated water over a long period[65].

There is evidence that cadmium toxicity may be reduced by simultaneous ingestion of some other metals. In animals, cobalt and selenium, as well as zinc, have been shown to have this modifying effect. As has been mentioned above, cadmium competes with zinc and other metals for binding sites on the protein metallothionein, indicating that metabolically-significant interactions occur between these elements[66]. Selenium has also been shown to have a protective effect against cadmium in experimental animals. A low intake of dietary copper may reduce tolerance to cadmium[67].

ANALYSIS OF FOODSTUFFS FOR CADMIUM

A wet digestion is the preferred sample preparation procedure for the determination of cadmium in foods. The Analytical Methods Sub-committee recommends use of a sulphuric acid, hydrogen peroxide wet digestion, with the usual precautions[68]. If dry ashing is used, the temperature must be kept below 500° C to prevent volatilisation losses[69].

It may be necessary to preconcentrate the element when foods very low in cadmium are being analysed. This may involve use of the complexing agent ammonium tetramethylene dithiocarbamate and extraction with 4-methylpentane-2-one (MIBK).

Flame AAS, using air-acetylene, has been shown to give accurate results, provided flame conditions are carefully controlled. Flameless AAS is preferable when concentrations of the element are low.

A procedure, using a graphite furnace connected to an instrument equipped with a deuterium lamp for background correction has been described by Jorhem and Slorach for the determination of cadmium in a variety of canned foods[70]. Graphite furnace AAS has also been used to determine cadmium in total diets, following complexing with ammonium pyrrolidine dithiocarbamate, by Dabeka and colleagues[71].

Cadmium has also been determined directly in milk, with very limited pre-analytical preparation apart from dilution and addition of a detergent solution, by Zeeman AAS as described by Narres and colleagues[72].

REFERENCES

1. BYCROFT, B.M., COLLER, S.A.W., DEACON, G.B., COLEMAN, D.J. and LAKE, P.S. (1982). Mercury contamination of the Lederderg River, Victoria, Australia, from an abandoned gold field. *Environ. Poll.*, 28, 135–47.

2. UNITED NATIONS ENVIRONMENT PROGRAMME; INTER-NATIONAL LABOUR ORGANISATION; WORLD HEALTH ORGAN-ISATION (1989). Mercury — environmental aspects. *Environ. Health Criteria* No. 86, 115 pages.

3. REILLY, C., SUDARMADJI, S. and SUGIHARTO, E. (1989). Mercury and arsenic in soil, water and food of the Dieng Plateau, Central Java, Indonesia. *Proc. Nutr. Soc. Australia*, 14, 117.

4. WORLD HEALTH ORGANISATION (1976). *Environmental Health Criteria: Mercury* (WHO, Geneva).

5. SWEDISH EXPERT GROUP (1971). *Methyl Mercury in Fish. A Toxicological–Epidemiological Appraisal of Risks. Nord. Hyg. Tdskr.* Suppl. 4.

6. WORKING GROUP ON MERCURY IN FISH (1980). Discussions on mercury limit in fish. *Australian Fisheries*, October 1980, 2–10.

7. NORDBERG, G.F., GOYER, R.A. and CLARKSON, T.W. (1985). Impact of effects of acid precipitation on toxicity of metals. *Environ. Health Perspect.*, 63, 169–80.

8. BENNET, B.G. (1984). Six most dangerous chemicals named. Monitoring and Assessment Research Centre, London, on behalf of UNEP/ILO/WHO International Program on Chemical Safety. *Sentinel*, 1, 3.

9. BERLIN, M.H., CLARKSON, T.W., FRIBERG, L.T. *et al.* (1963). Maximum allowable concentrations of mercury compounds. *Arch. Environ. Health.*, 6, 27–39.

10. BAKIR, F., DAMLUJI, S.F., AMIN-ZAK, L. *et al.* (1973). Methyl mercury poisoning in Iraq. *Science*, 181, 230–42.

11. STOEWSAND, G.S., BACHE, C.A. and LISK, D.J. (1974). Dietary selenium protection of methylmercury intoxication of Japanese quail. *Bull. Environ. Contam. Toxicol.*, 11, 152–6.

12. FURR, A.K., KELLY, W.C., BACHE, C.A., GUTENMANN, W.H. and LISK, D.J. (1986). Multielement uptake by crops grown in pots on municipal sludge-amended soil. *J. Agric. Food. Chem.*, 24, 889–93.

13. BOUQUIAUX, J. (1974). *CEC European Symposium on the Problems of Contamination of Man and his Environment by Mercury and Cadmium* (CID, Luxembourg).

14. OEHME, F.W. (1978). *Toxicity of Heavy Metals in the Environment* (M. Dekker, New York).

15. VOS, G., HOVENS, J.P.C. and DELFT, W.V. (1987). Arsenic, cadmium, lead and mercury in meat, livers and kidneys of cattle slaughtered in The Netherlands during 1980–1985. *Food Add. Contam.*, 4, 73–88.

16. MOHARRAM, Y.G., MOUSTAFA, E.K., EL-SOKKARY, A. and ATTIA, M.A. (1987). Mercury content in some marine fish from the Alexandria coast. *Nahrung*, 31, 899–904.

17. NATIONAL HEALTH AND MEDICAL RESEARCH COUNCIL (1987). *Food Standards Code* (Australian Government Publishing Service, Canberra, ACT).

18. CAPON, C.J. (1981). Mercury and selenium content and chemical form in vegetable crops grown on sludge-amended soil. *Arch. Environ. Contam. Toxicol.*, 10, 673–89.

19. SILLEN, L.G. (1963). *Sven. Kem. Tidskr.*, 75, 161–7.

20. MINISTRY OF AGRICULTURE, FISHERIES AND FOOD (1971). *Survey*

on *Mercury in Food*, and (1973) *Supplementary Report on Mercury in Food* (HMSO, London).
21. CAPON, C.J. and SMITH, J.C. (1982). Chemical form and distribution of mercury and selenium in edible seafood. *J. Anal. Toxicol.*, 6, 10–21.
22. HARADA, M. (1978). Methyl mercury poisoning due to environmental contamination ('Minamata Disease'), in *Toxicity of Heavy Metals in the Environment*, Part 1, ed. OEHME, F.W., 261–72 (Dekker, New York).
23. SKERFVING, S. (1974). Methylmercury exposure, mercury levels in blood and hair, and health status of Swedes consuming contaminated fish. *Toxicol.*, 2, 3–23.
24. TOMLINSON, G.H. (1979). Acid precipitation and mercury in Canadian lakes and fish. *Scientific papers from the Public Meeting on Acid Rain Precipitation*, May 1978, Lake Placid, New York (New York State Assembly, Albany, NY).
25. TURNER, M.D., MARSH, D.A., SMITH, J.C., ENGLIS, J., CLARKSON, T.W. *et al.* (1980). Methylmercury in populations eating large quantities of marine fish. *Arch. Environ. Health*, 35, 367–78.
26. CLARKSON, T.W. (1984). Mercury, in *Changing Metal Cycles and Human Health*, ed. NRIAGU, J.O., 285–309 (SpringerVerlag, Berlin).
27. MAGOS, L. (1975). Methylmercury poisoning. *Brit. Med. Bull.*, 31, 241.
28. MAGOS, L. (1971). Selective atomic absorption determination of inorganic mercury and methylmercury in undigested biological samples. *Analyst*, 96, 847–53.
29. CAPON, C.J. and SMITH, J.C. (1977). Gas-chromatographic determination of inorganic mercury and organomercurials in biological materials. *Anal. Chem.*, 49, 365–71.
30. SJOSTRAND, B. (1964). Simultaneous determination of mercury and arsenic in biological and organic materials by activation analysis. *Anal. Chem.*, 36, 814–19.
31. TANNER, J.T. and FORBES, W.S. (1975). Mercury in total diets. *Anal. Chim. Acta.*, 74, 17–23.
32. VOS, G., HOVENS, J.P.C. and DELFT, W.V. (1987). Arsenic, cadmium, lead and mercury in meat, livers and kidneys of cattle slaughtered in The Netherlands during 1980–85. *Food Add. Contam.*, 4, 73–88.
33. US ENVIRONMENTAL PROTECTION AGENCY (1975). *Scientific and Technical Assessment Report on Cadmium*, E.P.A. 600/6–75–003 (US Government Printing Office, Washington, DC).
34. NATIONAL HEALTH AND MEDICAL RESEARCH COUNCIL (1978). *Report on Revised Standards for Metals in Food.* (NHMRC, Canberra, ACT).
35. DABEKA, R.W., MCKENZIE, A.D. and LACROIX, G.M.A. (1987). Dietary intakes of lead, cadmium, arsenic and fluoride by Canadian adults: a 24-hour duplicate diet study. *Food Add. Contam.*, 4, 89–102.
36. VREMAN, K., VAN DER VEEN, N. G., VAN DER MOLEN, E. J. and DE RUG, W.G. (1986). Transfer of cadmium, lead, mercury and arsenic from feed into milk and various tissues of dairy cows: chemical and pathological data. *Netherl. J. Agric. Sc.*, 34, 129–44.
37. FITZGERALD, P.R., PETERSON, J. and LUE-HING, C. (1985). Heavy metals in tissues of cattle exposed to sludge-treated pastures for eight years. *Am. J. Vet. Res.*, 46, 703–7.
38. HAMMER, D.I., FINKLEA, J.F., CREASON, J.P. *et al.* (1971). Cadmium

exposure and human health effects, in *Trace Substances in Environmental Health*, ed. HEMPHILL, D.D., 269–87 (Uni. Missouri Press, Columbia, Mo.).

39. ASAMI, M.O. (1984). Pollution of soils by cadmium, in *Changing Metal Cycles and Human Health*, ed. NRIAGU, J.O., 95–111 (Springer Verlag, Berlin).

40. RAYMENT, G.E., BEST, E.K. and HAMILTON, D.J. (1989). Cadmium in fertilisers and soil amendments. *Chemistry International Conference*, Brisbane, 28 August–2 September, Royal Australian Chemical Institute.

41. KJELLSTRÖM, T., LIND, B., LINNMAN, L. and ELINDER, C.G. (1975). Variations of cadmium content in Swedish wheat and barley. *Arch. Environ. Health*, 30, 321–8.

42. JOHN, M.K. (1973). Cadmium uptake by eight food crops as influenced by various soil levels of cadmium. *Environ. Pollut.*, 4, 7–15.

43. BLANCH, K. (1990). *The Sunday Mail* (*Brisbane, Queensland*), 25 February, 5.

44. LISK, D. (1972). Trace elements in soils, plants and animals. *Adv. Agron.*, 24, 267–320.

45. LOUEKARI, K. and SALMINEN, S. (1986). Intake of heavy metals from foods in Finland, West Germany and Japan. *Food Add. Contam.*, 3, 355–62.

46. DABEKA, R.W. and MCKENZIE, A.D. (1988). Lead and cadmium levels in commercial infant foods and dietary intake by infants 0–1 year old. *Food Add. Contam.*, 5, 333–42.

47. WORLD HEALTH ORGANISATION (1963). *International Standards for Drinking Waters*, 2nd edn (WHO, Geneva).

48. LYON, T.D.B. and LENIHAN, J.M.A. (1977). Kitchen boilers as source of lead and cadmium. *The Lancet*, 1, 423.

49. ROSMAN, K.J.R., HOSIE, D.J. and DE LAETER, J.R. (1977). The cadmium content of drinking water in Western Australia. *Search*, 8, 85–6.

50. DE LAETER, J.R., WARE, L.J.. TAYLOR, K.R. and ROSMAN, K.J.R. (1976). The cadmium content of rural tank water in Western Australia. *Search*, 7, 444–5.

51. HOFFMAN, C.M. (1968). Trace metals in illicit alcoholic beverages. *JAOAC*, 51, 580–6.

52. DABEKA, R.W., MCKENZIE, A.D. and LACROIX, G.M.A. (1987). Daily intake of lead, cadmium, arsenic and fluoride by Canadian adults: a 24-hour duplicate diet study. *Food Add. Contam.*, 4, 89–102.

53. CODEX ALIMENTARIUS COMMISSION (1984). *Contaminants. Joint FAO/WHO Food Standards Program. Codex Alimentarius*, Vol. XVII, 1st edn (FAO, Rome).

54. OSTERGAARD, K. (1977). Cadmium in cigarettes. *Acta Med. Scand.*, 202, 193–7.

55. NWANKWO, J.N., ELINDER, C.G., PISCATOR, M. and LIND, B. (1977). Cadmium in Zambian cigarettes: an interlaboratory comparison in analysis. *Zambian J. Sc. Technol.*, 2, 1–4.

56. WORLD HEALTH ORGANISATION (1973). *Trace Elements in Human Nutrition. Tech. Rep. Ser. No. 532*, 41–3 (WHO, Geneva).

57. HAMMER, D.I., FINKLEA, J.F., CREASON, J.P. *et al.* (1971). Cadmium exposure and human health effects, in *Trace Substances in Environmental Health*, ed. HEMPILL, D.D., 269–88 (Uni. of Missouri, Columbia, Mo.).

58. GROSS, S.B., YEAGAR, D.W. and MIDDENDORF, M.S. (1976). Cadmium in liver, kidney and hair of humans, foetal through old age. *J. Toxicol. Environ. Health*, 2, 153–67.

59. FRIBERG, L. and VAHTER, M. (1983). Assessment of exposure to lead and cadmium through biological monitoring of a UNEP/WHO global study. *Environ. Res.*, 30, 95–123.

60. LAUWERYS, R.R. (1980). Significance of levels of cadmium in blood and in urine as measures of cadmium exposure, *Report on Seminar on Occupational Exposure to Cadmium*, 57–65 (Cadmium Assoc., London).

61. ROBINSON, J.W. and WEISS, S. (1980). The direct determination of cadmium in hair using carbon rod AAS. *J. Environ. Sci. Health*, A15, 663–7.

62. MORGAN, H. and SHERLOCK, J.C. (1984). Cadmium intake and cadmium accumulation in the human kidney. *Food Add. Contam.*, 1, 45–51.

63. FRIBERG, L., PISCATOR, M., NORBERG, G. and KJELLSTROM, T. (1974). *Cadmium in the Environment*, 2nd edn (CRC Press, Cleveland, Ohio).

64. BROWNING, E. (1969). *Toxicity of Industrial Metals* (Butterworths, London).

65. PISCATOR, M. (1981). *Carcinogenicity of Cadmium — Review. Third Int. Conf. on Cadmium*, Miami, Florida, 3–5 February.

66. SANDSTEAD, H.H. and KLEVAY, L.M. (1975). Cadmium–zinc interactions: implications for health. *Geol. Soc. Am. Spec. Pap.* 155, 73–83.

67. FRIBERG, L., PISCATOR, M. and NORBERG, G. (1971). *Cadmium in the Environment* (CRC Press, Cleveland, Ohio).

68. ANALYTICAL METHODS SUBCOMMITTEE (1975). Sample preparation. *Analyst*, 100, 761–3.

69. KJELLSTROM, T., LIND, B., LINNMAN, L. and ELINDER, C.G. (1975). Variations of cadmium concentration in Swedish wheat and barley. *Arch. Environ. Health*, 30, 321–8.

70. JORHEM, L. and SLORACH, S. (1987). Lead, chromium, tin, iron and cadmium in foods in welded cans. *Food Add. Contam.*, 4, 309–16.

71. DABEKA, R.W., MCKENZIE, A.D. and LACROIX, G.M.A. (1987). Dietary intakes of lead, cadmium, arsenic and fluoride by Canadian adults: a 24-hour duplicate diet study. *Food Add. Contam.*, 4, 89–102.

72. NARRES, H.D., MOHL, C. and STOEPPLER, M. (1985). Metal analysis in difficult materials with platform furnace Zeeman-atomic absorption spectroscopy 2. Direct determination of cadmium and lead in milk. *Z. Lebensm. Unters. Forsch.*, 181, 111–16.

CHAPTER 7

THE TOXIC METALLOIDS: ARSENIC, ANTIMONY AND SELENIUM

These three metalloids, chemically and physically lying between metals and non-metals, have had a long history in human affairs and have considerable interest for both the toxicologist and the nutritionist. All three figure in the Food Standards of many countries under Metal Contaminants in Food regulations.

Arsenic has been traditionally associated with homicide and the forensic scientist. No one would take lightly the presence of this element in food or drink. Yet arsenic is almost universally distributed in plants and animals, and daily intake, even in the absence of pollution, may be as high as 0.5 mg in all of us. It may even be that this poisonous element is essential for human life.

Selenium was for long better known for its toxicity than for any other quality. Farmers, whose cattle died from 'blind staggers' from eating certain plants, had no idea that the toxin in the fodder was, in fact, a very important nutrient for man and animals. Today the element is the centre of much research and is believed by some to offer hope in combating free radicals, the ravagers of cells and possible causes of cancer.

Antimony, though also ancient and long used by mankind, unlike the other two metalloids, is not believed to be a nutrient. It is toxic and occurs widely in many foodstuffs. As with arsenic and selenium, the chemical form in which the element occurs in food governs its toxicity.

ARSENIC

CHEMICAL AND PHYSICAL PROPERTIES

Arsenic is one of the Group V elements, which includes nitrogen, phosphorus, antimony and bismuth. Its atomic number is 33, and it has a density of 5.7. Its crystalline form is greyish in colour. While its melting point is 814° C, it has a boiling point of 613° C, at which temperature the element sublimes. This property is the basis of the traditional Marsh's test for arsenic.

The chemistry of arsenic is similar to that of phosphorus in many respects. Arsenic has oxidation states of -3, 0, and 3.5. Its two most common inorganic compounds are the trioxide (As_2O_3, white arsenic) and the pentoxide (As_2O_5). Other important compounds are arsenic trichloride and the various arsenates, such as lead arsenate, copper aceto-arsenate and the gaseous hydride arsine (AsH_3). Arsenic also forms organic compounds of interest, including arsanilic acid, and dimethylarsenilic acid.

The toxicity of arsenic and its compounds is related to the chemical form of the element. The inorganic compounds are the most toxic, followed by the organic arsenicals and finally by the gas, arsine[1].

PRODUCTION AND USES

Arsenic occurs widely in the earth's crust. It is found in soils, many waters and almost all plant and animal tissues. However, no major commercially-exploitable ore is known and the element is obtained as a by-product of the production of copper, lead or some other metal. It is extracted from flue gases, where it has collected after vaporisation during the smelting of the other metals. This flue dust is purified in an 'arsenic kitchen' to a mixture of about 97 per cent arsenic trioxide and 3 per cent oxides of other elements, of which the most important is antimony trioxide. World production of arsenic trioxide is about 60 000 tonnes per annum.

Arsenic has many applications in the metallurgical industry, not as a pure element but in alloys. It has the property of hardening and giving heat resistance to steel.

The principal use of arsenic is in the chemical industry and there it is widely employed in pharmaceuticals, agricultural chemicals, preservatives and related compounds. In fact, it is such uses that have conferred on arsenic the dubious honour of being second only to lead as a toxicant in the farm and in the household[1].

Agricultural and related substances using arsenic include herbicides, fungicides, wood preservatives of several kinds, insecticides, rodenticides and sheep-dips. These arsenic-containing products were once far more widely used than they are today when restrictions have been placed on them because of their toxicity. Unfortunately, arsenic is a persistent poison and, even after many years, some streams and locations on farms remain toxic because they were once exposed to these chemicals. The most important of the agricultural chemicals of this type include lead arsenate, copper arsenate, copper aceto-arsenate (Paris Green), sodium arsenate (Wolman salts) and cacodylic acid, all once well known to farmers.

It was because of concern at the possibility of forward transmission of arsenic in the human food chain that the use of many of these substances is now restricted or totally banned. Lead arsenate, for instance, formerly used extensively as an insecticide on tobacco, resulted in a higher than average intake of the element by smokers. Similarly, the use of arsenical sprays to prevent infestation by insects of apples, resulted in contamination of fruit and of cider. The use of arsenic in sheep-dip has resulted in many cases of stock and even of human poisoning on farms. Other agricultural uses of arsenic compounds include the incorporation of arsanilic acid into pig and poultry feed as growth-promoters.

The chemical industry uses arsenic in the manufacture of dyestuffs and in glass and enamels. In former times arsenic was widely used in medicine. Until relatively recently a patent medicine in many pharmacies was Fowler's solution, an arsenic-containing preparation for treatment of skin problems and, even taken internally as a tonic for preventing anaemia. A report published in 1965 on a large group of patients who had used Fowler's solution over many years, found that a significantly high proportion of them had developed skin cancer[2]. Organic arsenical compounds such as Salvarsan have been extensively used for treating syphilis, but are now superseded. Certain arsenical preparations, including some amoebicides, are still prescribed today. In the past it was probably the less restricted availability of arsenic-containing pharmaceutical compounds that made arsenic, especially in the form of its trioxide, one of the most common homicidal and suicidal poisons for many centuries.

ARSENIC IN FOOD AND BEVERAGES

Because of its wide distribution in the environment and its use in agriculture, arsenic is present in most human foods. It is usually present in very low concentrations of less than 0.5 mg/kg and rarely exceeds 1 mg/kg, except in seafoods, including fish and seaweeds.

Soil usually contains between 1 and 40 mg/kg in the absence of industrial or agricultural contamination. The use of sewage sludge as a soil improver can increase these levels considerably. The arsenic may come from its use in some detergents. Phosphate fertiliser may contain appreciable levels of arsenic[3]. Uptake of the element from soil by plants has been found to depend not only on soil concentration but also on the species of plant. Some plants are, in fact, capable of accumulating arsenic in their tissues.

A UK study, based on food collected in a market basket survey in 1978[4] reported that the mean value of arsenic in all foods and beverages investigated, except for fish, was less than 0.02 mg/kg and that the estimated total daily intake of an adult was less than 81 μg. Fish contained 2.71 mg/kg and contributed 54 μg of that daily intake.

A recent Canadian study[5], utilising improved analytical techniques, found the following levels of arsenic in different food groups (expressed as mean values, μg/kg, with ranges in parentheses): cereals 8.6 (0.71–61); dairy products 2.58 (<0.6–11.3); starchy vegetables 13.69 (<4–81.9); other vegetables 2.60 (<0.6–8.3); meat/fish 60.1 (<4–625). Most of the daily intake of arsenic was from three food groups, 18.1 per cent of total intake from cereals and cereal products, 14.9 per cent from starchy vegetables and 32.1 per cent from the meat/fish group. Individual foodstuffs in the latter group were found to contain high levels of arsenic, with 2.3 mg/kg in clams, 1.0 mg/kg in salmon and 0.39 mg/kg in tuna.

Daily intake of arsenic found in the Canadian study ranged from 2.6 to 101 μg/day, with a mean of 16.7 μg. This is equivalent to an average intake of 0.26 μg/kg/day. The highest intake recorded for a participant in the study was by a man who obtained 79 per cent of his total arsenic from fish and meat. The next highest intake was appreciably lower at 33.7 μg/day.

The Canadian intake figures were lower than those reported from several studies in other countries. Figures of 55 μg/day have been reported for New Zealand[6], 62 μg/day for the US[7], and 89 μg/day for the UK[8]. However, closer to the Canadian figures are the reports of 15–45 μg/day for Sweden[9], and 12 μg/day in Belgium[10].

Dabeka and his colleagues comment on their findings for Canadian adults that the average arsenic intake of 0.26 μg/kg/day was lower than the FAO/WHO maximum acceptable load for humans of 2 μg/kg[11]. They also note that only one participant in their study, the heavy meat and fish eater whose daily intake was 101 μg, equivalent to 1.7 μg of arsenic per kg body weight each day, approached the FAO/WHO figure.

Arsenic in Water

Arsenic is detectable in almost all potable waters. The usual levels of concentration ranges from 0 to 0.2 mg/litre with a mean of 0.5 μg/litre[12]. US Federal Regulations for drinking water set a maximum of 0.01 mg/litre[13]. Particular sources, such as spas and hot springs, may have considerably higher concentrations of arsenic, depending on the nature of the underlying rock and sources of the water. What has been described as 'regional endemic chronic arsenicism' has been attributed to domestic use of naturally-contaminated waters. Levels of 1–4 mg of arsenic trioxide per litre have been detected in such water in the Argentine and Chile. The consumption of well-water containing 0.6 mg/litre in parts of Taiwan has been associated with chronic arsenic poisoning[14].

Arsenic in Fish, Seafoods

As has been seen above, fish and other seafoods can accumulate high levels of arsenic and contribute significantly to daily intake of the element. Accumulation can be particularly high in shellfish, especially if living in polluted waters. Levels of up to 26 mg/kg have been found in crab, and 40 mg/kg in shrimps tested in the UK[4] and, in one case, 170 mg/kg in prawns.

Free-swimming fish do not generally contain as much arsenic as do shellfish. Figures from the UK survey found a range of 1.0–6.0 mg/kg in haddock and <0.5–2.4 in herring. However, flounder and sole from a polluted estuary had ranges of <0.2–34, and 0.5–24 mg/kg respectively. An Australian study of marine commercially used fish found up to 4.4 mg/kg of arsenic in one species, while 21 per cent of those tested had levels above the Australian maximum permitted level of 1.0 mg/kg[15].

It is appropriate to mention here that food regulations in Australia, and in some other countries, while setting a maximum permitted level of arsenic in 'fish, crustaceans and molluscs', apply this level to 'inorganic arsenic only' and do not take into account levels of organic arsenic in the foodstuffs[16]. This approach recognises that, unlike the situation with regard to lead, mercury and certain other elements, the organic form of arsenic is not believed to be toxic, in contrast to the highly toxic inorganic forms. As a WHO Expert Committee noted, arsenic in muscle tissue mostly 'remains organically bound and studies in man using isotopically labelled organoarsenic derivatives present in chicken liver paté have revealed that they are rapidly excreted by man without significant retention of elemental arsenic'[17]. Monier-Williams, in his pioneering study of trace elements in food, had already noted that arsenic present in marine organisms is in the form of an organic complex which is readily excreted by humans and is of low toxicity[18].

Arsenic from Industrial Pollution

Industrial pollution and accidental contamination can result in higher than normal levels of arsenic in foodstuffs and beverages. Environmental contamination around a coal-burning power-station in Novaky, Czechoslovakia, was shown to cause considerably increased levels of arsenic in locally-grown crops and in water[19]. The coal used at the power-station contained more than 200 times as much arsenic as do US coals, in which arsenic concentration is on average 5 g/tonne. The Czechoslovakian plant emitted up to 1 tonne of arsenic from its smokestacks each day. Drinking water in the surrounding countryside contained 0.07 mg/litre, and surface water 0.21 mg/litre of arsenic. It was believed that chronic arsenic poisoning occurred in long-term residents of Novaky.

Chronic arsenic poisoning, in animals and possibly humans, has been alleged to have occurred in the UK in the vicinity of coal-fired brick kilns[20]. Arsenic emissions from metal smelters have been reported from the USA and Sweden, with, in the latter case, up to 10 tonnes a day being released[21].

Accidental Contamination of Foods by Arsenic

Accidental contamination of foods with arsenic has resulted in a number of serious cases of poisoning. An early, and fully-recorded case which resulted in the setting up of a Royal Commission, occurred in the UK in 1900, when arsenic-contaminated beer poisoned more than 6000 drinkers in Lancashire and Staffordshire, 70 of them fatally. The Commission found that the outbreak resulted from the use of arsenical pyrites to make the sulphuric acid which was used to hydrolyse starch to make the glucose used in fermentation[4].

An even more serious accident occurred in Japan 55 years later when more than 12000 infants fed on a formula which had been contaminated with arsenic were poisoned. One hundred and twenty of them died. The arsenic was traced to the sodium phosphate used to stabilise the formula. This chemical had been produced as a by-product during refining of aluminium from bauxite which, unsuspectedly, contained substantial amounts of arsenic[22].

The use of arsenicals as pesticides in vineyards has resulted in some cases of arsenic poisoning of wine drinkers. Increased dietary intake has also occurred from residual arsenic in horticultural chemicals used on vegetables. Though there has been a decrease in the use of arsenicals in agriculture in recent years, they are still used extensively on certain non-food crops, including cotton. Cotton seed by-products are used as human foodstuffs and there is a danger that this use can be a pathway for arsenic into the diet[23]. Because of such dangers, several countries set tolerance limits to arsenic residues in foodstuffs. This is 3.5 mg/kg in the US[24].

UPTAKE AND EXCRETION OF ARSENIC BY THE BODY

Both trivalent and pentavalent arsenic are easily absorbed from food in the gastro-intestinal tract. Actual quantities absorbed appear to depend on the chemical form of the element present. The absorbed arsenic is quickly transported to all organs and tissues as a complex, probably with alpha-globulin proteins. After 24 h, concentrations in the organs generally begin to decrease, as arsenic is lost from the body. This is not the case in skin where there can be an increase over several days. Accumulation takes place in skin, nails and hair, and, to some extent, in bone and muscle.

Total body levels of arsenic have been reported to be between 14 and 20 mg[3]. The biological half-time is believed to be short, between 10 and 30 h for both inorganic and organic arsenic. Excretion is mainly in the urine.

Levels of arsenic in blood and urine have been used to measure exposure to the element. Hair has also been used, especially in investigations of homicides. In some cases it has been possible to derive a chronological record of ingestion of arsenic over time by this means. Hair arsenic levels from 5 to 700 mg/kg have been found in cases of acute arsenic poisoning[25].

METABOLISM AND TOXIC EFFECTS OF ARSENIC

Arsenic can cause both chronic and acute poisoning. Arsenic trioxide is a common cause of the latter. A fatal dose is of the order of 70–180 mg.

The pentavalent form is less toxic than the trivalent form of the element. Arsenic, in any chemical form, is a general protoplasmic poison. It binds organic sulphydryl groups and thus inhibits the action of enzymes, especially those concerned with cellular metabolism and respiration. The principal pharmacological effects are dilation and increased permeability of capillaries, especially in the intestines.

Chronic arsenic poisoning causes loss of appetite, leading to weight loss. This is followed by gastro-intestinal disturbances, peripheral neuritis, conjunctivitis, and skin problems including hyperkeratosis and melanosis. This latter darkening of the skin is characteristic of prolonged exposure to arsenic and may be a factor in the development of skin cancer. It has been suggested that arsenic, in some circumstances, but not necessarily in the diet, may be carcinogenic[26].

Interaction between Arsenic and Other Elements

It has been reported that arsenic can counteract the toxicity of an excessive intake of selenium[27]. The element has been added to the feed of poultry and cattle in areas of naturally high selenium for this purpose. Sodium selenite, injected simultaneously with sodium arsenite, which is teratogenic in animals, has been shown to prevent malformation of foetuses in animals. Interactions between arsenic and cadmium have also been demonstrated. It is possible that common modes of action of the different elements on cells and tissues, especially through interaction with protein −SH groups, account for some of these observations.

ANALYSIS OF FOODSTUFFS FOR ARSENIC

The traditional Marsh test for arsenic, which relied on the volatility of the element, was qualitative rather than quantitative. It was well-known to the forensic scientist, as well as the writer of detective stories in former years. It has been improved and generally replaced by accurate, instrumental quantitative procedures.

A spectrophotometric method for the determination is still employed by some investigators, with excellent results. Bebbington and his colleagues used this method in their study of arsenic in Australian fish[15]. They digested their samples with nitric and perchloric acids, making sure that they had an excess of nitric acid to prevent reduction of As^{5+} to the more volatile As^{3+}. The residue remaining after digestion and evaporation of the acids was dissolved in water and assayed by measuring the red colour at 535 nm produced by the reaction of arsine with silver diethyldithio-carbamate. Arsine was generated by reduction of inorganic arsenic in acid solution in a Guzeit generator. A detection limit of 0.1 mg/kg was obtained, with recoveries of 90–95 per cent.

Dry ashing, using alkaline magnesium nitrate, may be used in sample preparation for atomic absorption and arsine generation techniques. Simple dry ashing is unsuitable since it causes loss of volatile arsenic compounds. Wet digestion mixtures, using nitric/sulphuric mixtures, with or without perchloric acid, are suitable for graphite furnace AAS. Solvent extraction preconcentration techniques using, for example, ammonium pyrrolidine dithiocarbamate, are recommended when expected concentrations of arsenic are low[28].

Graphite furnace AAS, using either deuterium or Zeeman background correction, have been found to be suitable for the determination of low levels of arsenic in foods and other materials[29]. However, pre-concentration, by co-precipitation or dithiocarbamate extraction, is normally necessary. Detection limits of 0.3 μg/kg were found using GFAAS, following preconcentration[28].

Hydride-generation has been noted as the best current technique for nanogram amounts of arsenic by Tsalev[30]. This technique is based on the conversion of arsenic into its gaseous hydride, arsine. Commercially produced arsine generators are available and are used in conjunction with standard atomic absorption spectrophotometers. In the method arsenic compounds in the digested sample are reduced to the trivalent state in a reaction vessel and AsH_3 is generated. This is passed into a nitrogen–hydrogen flame, supported by entrained air. Absorption is measured at 193.7 nm. Background correction by deuterium lamp or Zeeman effect is required to overcome interference.

Sensitive AES procedures with plasma-source excitation have been developed. Further developments in ICP determination of arsenic are expected and, where the instrumentation is available, are likely to replace both AA and hydride generation techniques because of their better sensitivity[30].

ANTIMONY

CHEMICAL AND PHYSICAL PROPERTIES

Antimony, which has been given the chemical symbol Sb from its ancient classical name of *stibium*, has an atomic weight of 121.8 and is number 51 in the Periodic Table of the elements. It is a fairly heavy element with a density of 6.7. Though closely related chemically to arsenic, its physical properties are more metallic. It melts at 631° C and vaporises at 1750° C. The crystalline form of antimony is silver-white, hexagonal in shape and metal-like in appearance. Oxidation states are 3 and 5. Among its most important compounds are the tri- and pentoxides, the trichloride, and the tri- and pentasulphides. The trihydride, stibine, is a gas. Several organic compounds are also known.

PRODUCTION AND USES

The chief natural source of antimony is the ore stibnite, which is mined in China and, to a lesser extent, in Central and South America. Antimony is also obtained as a by-product of the refining of other metals.

When combined with other metals, antimony hardens them and is used in alloys for this purpose. It is combined with lead for making storage-battery plates, and with tin and copper in bearings and other machine parts. Formerly one of its most important uses was as an alloy, combined with lead and tin for making printer's type. When cast the resulting type had clean, sharp edges and was ideal for printing. Other uses of the metal is for the manufacture of solder, ammunition and electrical cable coverings.

Antimony also has wide use in the chemical industry to produce fireproofing chemicals, as well as in the manufacture of paints and lacquers, rubber, glazes and pigments for ceramic and glass manufacture. The pharmaceutical industry uses some antimony in preparations for the treatment of parasitic infestations. One of these, which is still widely used in the tropics, is tartar emetic. In ancient times the metal was valued as a cosmetic, especially as an eye-shadow.

METABOLISM AND BIOLOGICAL EFFECTS OF ANTIMONY

Physical contact with antimony and its compounds, as well as fumes and dust of the element are capable of causing dermatitis, conjunctivitis and nasal septum ulceration. The hydride, stibine, is very dangerous and has caused death when inhaled[31]. Most of the known poisoning incidents have been due to industrial exposure, but a few cases resulting from ingestion of antimony have been reported. However, ingested antimony has apparently a low inherent toxicity[32]. In most cases the source of the antimony has been soft drinks which had been stored in enamelled containers[33]. Symptoms of poisoning are colic, nausea, weakness and collapse with slow or irregular respiration and a lowered body temperature.

Not very much is known about the uptake and behaviour of antimony in the human body. From animal experiments it appears that about 15 per cent of ingested antimony is absorbed in the gut. The element is concentrated in organs, including liver, kidney and skin. Excretion is rapid initially, but there may be a long-term component retained in the body. There is no evidence that antimony is an essential trace nutrient either for humans or other animals.

ANTIMONY IN FOOD AND BEVERAGES

Antimony has been included among the elements whose levels in foodstuffs are restricted by food regulations in several countries because of its recognised toxicity. Australian Food Standards, for instance, set a maximum permitted concentration in beverages of 0.15 mg/kg and 1.5 mg/kg in all other foods[34].

Little is known about daily intakes or even about usual levels in foods in most countries. It is believed to be between 0.25 and 1.25 mg/day for children in the USA[35]. Levels in individual foodstuffs are lacking. Concentrations of 0.1 to 0.2 μg/litre have been reported for both river and seawater[36]. The US Environmental Protection Agency has recommended a limit, normally not exceeding 0.1 mg/litre, for antimony in drinking water.

Monier-Williams, in another of his pioneering reports on trace elements in foods which has been referred to above[33], reported on high levels of antimony in certain foodstuffs due to contamination. Many of these resulted from the use, for cooking or storage of food, of containers glazed with antimony-containing enamel. He found that as much as 100 mg/litre of antimony could be dissolved by a dilute food acid solution from some enamelware. He noted a case of poisoning involving 56 persons who drank lemonade made in an enamelled bucket.

ANALYSIS OF FOODSTUFFS FOR ANTIMONY

Because of low concentrations in foods and beverages, it is usually necessary to pre-concentrate antimony by solvent extraction before analysis. Solid samples may be digested either by dry ashing, using magnesium nitrate, or wet digestion in an acid mixture.

Graphite furnace, with deuterium lamp or Zeeman background correction, can be used effectively for determination of even very low levels of the element. The presence of chloride in the digest can cause problems, however, and may have to be removed by back extraction into alkaline ammonium tartrate to ensure freedom from this interference.

The vapour generation technique, using the hydride stibine, is effective for solutions with low concentrations of the element. A procedure similar to that for arsenic is used, with appropriate instrument settings. There may be problems with stability of the hydride if certain elements are present in the solution[37].

SELENIUM

Though selenium itself was only discovered by Berzelius in the early nineteenth century, its toxicity to animals was recognised as long ago as 1295 when Marco Polo, travelling in western China, described what we now know as selenosis in horses[38]. This toxicity was confirmed in many later investigations. It occurred over a wide area of the semi-arid regions of the mid-US, where it poisoned horses and cattle, and was given the name 'alkali disease'. The earliest reports were of poisoning of cavalry horses fed on herbage growing in what came to be known as 'poison areas'[39]. Its symptoms were lameness, loss of hair, hoof cracking, blindness and paralysis. In horses it was sometimes called 'change hoof disease' and in cattle 'blind staggers'.

The disease was traced to the plants eaten by the animals. These were natural accumulators of selenium. In the US Mid-West the pasture plant *Astragalus racemosus*, was found to be able to accumulate up to 15 g/kg of selenium, more than enough to poison an animal. In Queensland, Australia, among several plants with similar properties, *Neptunia amplexicaulis* can accumulate up to more than 4 g/kg of the element. Browsing of this plant has resulted in acute selenosis in cattle and sheep[39].

Selenium poisoning of stock has been reported in many parts of the world besides China, USA and Australia. These include Mexico, Canada, Colombia, Israel, and Ireland. Selenosis is recognised as a real danger for farm animals under certain soil conditions. It was also recognised that the element represented a danger for humans who might eat vegetables or other foods with high concentrations of selenium.

It was a considerable surprise, therefore, when this highly toxic, and for some farmers at least, commercially disastrous substance, turned out to be an essential trace element, for animals as well as for humans. Indeed, as was found in the country where Marco Polo first came across its toxic trail, there are some regions in the world where people suffer from selenium toxicity, due to high environmental levels of the element, and others, in the same country, where the element is so scarce that selenium deficiency occurs. China is in the unhappy situation of having within its borders three endemic diseases, all associated with selenium, one of excess (selenosis) and the other two with deficiency (Keshan and Kashin-Beck). All three are serious conditions and have been the subject of extensive investigation in recent years.

As we shall see, selenium is, today, the subject of great interest, both as a nutrient and a toxin, and world-wide research is being carried out on the element.

CHEMICAL AND PHYSICAL PROPERTIES

Selenium occurs in Group VI of the Periodic Table, along with sulphur. The two elements, with tellurium and polonium, make up the sulphur family. Selenium has an atomic weight of 78.96 and is number 34 in the Table. Its density is 4.79. It occurs in a number of allotropic forms, one of which is metallic or grey selenium. This crystalline form contains parallel 'zigzag' chains of atoms and is the stable form at room temperature. It possesses the ability to conduct electricity more easily when it is exposed to light. It has also been shown that light generates a small electric current in selenium. These properties are the basis of the use of selenium in photoelectric cells.

Chemically selenium is close to sulphur, with many similar properties, though it is a weaker oxidising and reducing agent. It burns in oxygen to form the dioxide, which is a solid. It also combines directly with halogens and with many metals and non-metals. Its oxidation states are -2, 0, $+2$, $+4$ and $+6$. All, except for $+2$, are commonly found in nature.

Selenium in the $+6$ or selenate state is stable under both alkaline and oxidising conditions. In the $+4$ state the element occurs naturally as selenite. This tends to oxidise slowly under alkaline conditions to selenate. It is readily converted to elemental selenium, by reducing agents including ascorbic acid. In the -2 state selenium exists as hydrogen selenide and a number of selenides. H_2Se is a gas at room temperature.

A large number of Se analogues of organic sulphur compounds are known, such as selenomethionine and selencysteine. These can replace ordinary amino acids in protein molecules and may represent storage forms of the element in biological tissues.

These different inorganic and organic forms of selenium are of considerable significance in relation to the biological function and the toxicity of the element. There is evidence that the nutritional role of selenium depends on whether the element occurs in organic or inorganic form in foods[40].

PRODUCTION AND USES

Selenium is not abundant, though it is a widely distributed element in the earth's crust. Soils generally contain about 0.2 mg/kg. As has been noted above, it can also occur in much higher levels in certain areas. Coal in some parts of the world contains high concentrations of the element and can release it into the environment when burned. This is believed to be the cause of high levels of soil selenium in certain seleniferous areas of China[41]. It occurs in a number of ores of other elements, usually as selenide. Its principal partner is sulphur and the pure element is normally extracted from flue dusts produced during the combustion of metallic sulphide ores, including iron pyrites. Selenium is also obtained as a by-product from sulphuric acid production in lead chambers. Most of the world's supply comes from Canada and the USA, with smaller amounts from Zambia and some other countries.

Selenium has a number of important industrial uses, one of which has been mentioned already, the manufacture of photoelectric cells. The electronic industry uses the element in some quantity, in photocopying machines, switches, rectifiers and other related equipment.

The element is also of importance for the manufacture of certain stainless steels and copper alloys. Selenium improves the machinability of such alloys. It is also used in glass manufacture, in ceramics, pigments and in the vulcanising of rubber. The animal feed and the pharmaceutical industries also use selenium as feed and nutritional supplements and in other ways.

SELENIUM IN FOOD AND BEVERAGES

Information on levels of selenium in foods and diets is not readily available. Analytical problems as well as, until relatively recently, probably, limited interest among investigators, have been responsible for this lack of data. The situation is now rapidly changing.

It is becoming clear that there can be significant differences in concentrations of the element in foods and in dietary intake in different countries of the world. These differences are almost entirely due to the level of the element in agricultural soils in each country. This is seen in estimates of daily intakes of selenium in different countries. These, as summarised by Thomson and Robinson[42], are (expressed as $\mu g/day$): New Zealand (South Island) 28; USA 132.0; Canada 98.3–224.2; Japan 88.3; Venezuela 325.8; UK 60; Italy 12.75. To these may be added: China[38], selenosis region: 4.99 mg/day; high selenium, but non-selenosis region: 750 $\mu g/day$; selenium-deficient region (Keshan disease): 11 $\mu g/day$.

It is of significance, with regard to the above wide range of intakes, that only in China, in Enshi County, where selenosis occurs and in Keshan and some regions of Mongolia where selenium-deficiency diseases are endemic, are symptoms of either selenium deficiency or excess clearly observed in the population. Even in China, as in Chile, high levels of intake of selenium do not seem to be necessarily toxic.

Low levels of intake of selenium in the South Island of New Zealand, have been a matter of concern for many years. Extensive investigations have been carried out[43] which have shown that the underlying cause of low dietary intake, and consequent low blood levels in the island, is the very low soil concentration of the element. Yet, there are no significant clinical signs of cardiomyopathy or other indications of selenium deficiency disease. It is highly likely that deficiency and excess of selenium bring about clinical symptoms only when other nutritional and toxicological factors, such as a generally inadequate diet, or the presence of other toxins, including mycotoxins, are at work on the victims at the same time. This seems to be clearly the case with one of the selenium-deficiency diseases encountered in China, Kashin-Beck disease, which results in bone deformation[44]. The disease occurs in isolated and impoverished areas with a severe climate. Climatic and nutritional stress appear to be major factors, along with selenium deficiency, in the development of the disease.

It is of interest, as has been mentioned earlier, that health authorities in Finland took steps to change, on a national scale, levels of dietary intake of selenium in the belief that this would in the long term lead to improved health in the community, especially with regard to cancer. The change in intake levels was brought about by the use of selenium-enriched fertiliser. Within about 10 years, dietary intake of selenium had been improved from less than 50 to more than 100 μg/day[45]. Whether this has the hoped-for health effects is still to be seen.

Some of the problems regarding selenium levels in food have probably resulted from the use of less than adequate analytical instruments and techniques. Selenium is not an easy element to determine, but in recent years the availability of Zeeman AAS and updated spectrofluorometric analytical procedures have improved the situation. The availability of Certified Reference Materials has also been a factor in this improvement in reliability of analytical results. These improvements may, in fact, account for the fact that recent results indicate a narrower spread of concentration differences between diets in different countries.

A daily intake of 34.7 μg by infants in Germany has been reported by Lombeck[46]. A US study[47] found an intake of 58.8 μg/day in older children (4–7 years). A recent investigation by the author[48] found a value of 55.9 μg/day for children of the same age in Australia.

Actual levels of selenium in individual foods and food groups found in the recent Australian study are (expressed as mg/kg): cereals and cereal products: 0.01–0.31; meat, fish, eggs: 0.06–0.34; milk and dairy products: <0.001–0.11; vegetables and fruit: <0.001–0.022. These are very similar to data from the UK[49].

Another study by the same Australian group[50] found that a significant contribution to selenium intake in the diet of children was made by cereals (36 per cent of total) and dairy products (25 per cent), with meat and fish responsible for only 17 per cent. Fruit and vegetables together supplied less than 2 per cent. In contrast, Lombeck's German study of the diet of infants[46] found that fruit and vegetables supplied 20 per cent of intake. Whether this represents differences in levels of selenium in fruit and vegetables in the two countries, or, more probably, differences in patterns of food intake of infants and older children, is not clear. However, overall intake, as has been noted, is comparable in both diets.

Some other studies have found higher concentrations of selenium in fish and meat than have been reported from the UK or Australia. Canadian sea fish were reported to contain up to 0.9 mg/kg[51] and Norwegian lobster roe 4.43 mg/kg[52]. Another Australian study[53] of nine species of commercial fish did not find any with selenium levels over 0.8 mg/kg.

Nuts, and in particular Brazil nuts, are a rich source of selenium. The UK study[49] found individual Brazil nuts containing as much as 53 mg/kg of the element. Cashew, walnut and some peanuts were also relatively rich in the element.

Few countries have recommended daily intakes of selenium, or even of safe and adequate daily intake ranges. The Food and Nutrition Board of the US National Research Council recommends a range of 50–200 μg/day, with 10–40 for infants and 20–120 for children of 1–6 years of age[54]. Australia has recently introduced a Recommended Daily Intake of 50–80 μg/day for adult males, 50–70 μg/day (+ 10 in pregnancy and lactation) for women, 25–30 μg/day for children (1–7 years) and 10–15 μg/day for infants[55].

METABOLISM AND BIOLOGICAL EFFECTS OF SELENIUM

Selenium occurs in food mainly in the form of seleno-amino acids, such as selenomethionine. Inorganic selinite and selenate are used usually as nutritional supplements, for man as well as farm animals. About 80 per cent of organic selenium ingested in food appears to be absorbed. Absorption of the inorganic forms seems to be less efficient, but still high. There is some evidence that selenium in plant foods is more easily absorbed than that in meat or other animal products[56].

Selenium is absorbed mainly in the duodenum, caecum and colon. It is transported in the blood stream bound to protein. There does not appear to be any special accumulation in any organ or tissue. Excretion is mainly through the urine, though some may appear in faeces with bile. When high levels of selenium are ingested, excretion of a volatile metabolite of the element, dimethylselenide, may occur through the lungs. This substance accounts for the garlic-like odour detected on the breath of those suffering from selenium toxicity.

The only well-established metabolic role for selenium in the body is as a prostetic group in the enzyme glutathione peroxidase[57]. The enzyme occurs in the cytosol or fluid portion of cells as well as in their mitochondria. Glutathione peroxidase plays an important role as one of the free radical scavengers of the body. Free radicals are highly reactive forms of atoms or groups of atoms, including superoxide and hydrogen peroxide, which can cause severe oxidative damage to cells and their components. Polyunsaturated fatty acids within cell membranes are particularly sensitive to attack by free radicals[58]. These active atoms are believed by some to be implicated in such illness as coronary heart disease, cancer, kwashiorkor and general ageing of the body. Glutathione peroxidase functions as part of a multicomponent antioxidant defence system, which includes vitamin E. These two nutrients work together in an as yet not fully understood fashion.

Another role of selenium that is becoming better known is its ability to interact with heavy and other metals, including cadmium, mercury and silver. It appears to be able to counteract the toxic effects of these metals and has been used as an antidote against mercury and cadmium poisoning. Selenium has been shown to be effective against methyl mercury toxicity. The breakdown of methyl mercury in the body is believed to result in the formation of free radicals and it is suggested that selenium exerts an antioxidant effect against them[59].

Selenium toxicity is, as we have seen, well documented in farm animals. Human selenosis has also been known and studied in the Chinese seleniferous region of Enshi County[41]. There is some evidence, also, of diet-related selenium toxicity in residents of other seleniferous regions of the world. Symptoms similar to those observed in animals, including dermatitis, dizziness, brittle nails, gastric disturbances, hair loss and, more distinctive, garlic odour on the breath, have been reported in humans in seleniferous areas of South Dakota[60], and, more recently, from Venezuela and Chile[61]. There is also evidence that medication with selenium nutritional supplements, both self-selected and medically directed, has resulted in selenium toxicity[62].

Because of the danger of toxicity from ingestion of selenium, and the growing interest in use of the element both as a human and veterinary nutritional supplement and growth promotor, authorities in several countries have placed restrictions on levels permitted in foods. This is especially important since the margin between toxicity and requirement is relatively narrow[63]. In Australia a maximum permitted limit of 0.2 mg/kg in beverages and other liquid foods, 2.0 in edible offal and 1.0 in all other foods, has been established[64].

The US Food and Nutrition Board (FNB) recommend an upper limit of daily intake of 200 μg^{54}. Higher limits have been suggested by others on the grounds that natural exposure often exceeds the FNB limit, without clinical effects[65]. However, a cautious and probably more acceptable view by another expert is that 'on balance, prolonged intakes of selenium in any form above 600 μg/day would seem to be unwise and on current evidence appear to entail a real risk of chronic selenium toxicity. These levels would be unlikely to be exceeded by normal dietary means, as even deliberate selection of selenium-rich foodstuffs would probably not provide more than 500 μg/day'[55].

Selenium deficiency, in humans and animals, has been given wide publicity and been the subject of extensive investigation since the condition was first recognised. Selenium-responsive deficiency diseases recognised in farm and experimental animals include exudative diathesis and pancreatic fibrosis in poultry, *hepatosis dietetica* in pigs and white muscle disease in lambs, foals, calves, rabbits and even marsupials[66]. Symptoms of these various diseases include cardiomyopathy, muscular dystrophy, as well as necrosis of liver and other organs.

Deficiency symptoms of similar severity to those seen in animals have not been described in humans while Keshan disease, encountered in selenium-deficient parts of China, is characterised by cardiomyopathy[67]. In other countries with low soil levels of selenium, no clear clinical symptoms of selenium-related illness have been observed even when intake of the element was as low as those met with in China[68]. However, there is evidence that selenium deficiency resulting from prolonged total parenteral nutrition has led to muscular discomfort and cardiomyopathy similar to that of Keshan disease[69].

There have been several studies which indicated that selenium deficiency may be linked to cancer. These have been mainly of an epidemiological nature and have linked cancer incidence with environmental deficiencies of the element[70].

ANALYSIS OF FOODSTUFFS FOR SELENIUM

Several methods can be used for the determination of selenium in foods and beverages. Among those widely used are hydride generation and graphite furnace AAS, with deuterium lamp or Zeeman mode background correction, and spectrofluorometry.

Where available, neutron activation analysis has proved to be an excellent method of determining the element in biological samples. A simple and rapid NAA method for selenium in human serum has been reported by McOrist and colleagues[71]. A similar method has been described for hair and other human tissues[72].

For AAS, samples should be carefully digested, in an acid mixture. Dry ashing will cause losses, as will wet digestion if charring is allowed to occur.

For the hydride generation procedure, gentle, wet digestion with a nitric–perchloric–sulphuric acids mixture, followed by fuming to remove nitric acid completely, is necessary.

Excellent recoveries and sensitivities are reported with the spectro-fluorimetric procedure. This involves digestion in a nitric–perchloric acids mixture to which hydrogen peroxide is added to reduce selenium to the tetravalent state. The selenium is then complexed with 2,3-diaminonaph-thalene (DAN) and extracted into cyclohexane[30].

REFERENCES

1. BUCK, W.B. (1978). Toxicity of inorganic and aliphatic organic arsenicals, in *Toxicity of Heavy Metals in the Environment*, ed. OEHME, F.W, 357–69 (M. Dekker, New York).
2. FIERZ, U. (1965). Skin cancer and arsenic-containing pharmaceuticals. *Dermatol.*, 131, 41–58.
3. SCHROEDER, H.A. and BALASSA, J.J. (1966). Abnormal trace metals in man: arsenic. *J. Chron. Dis.*, 19, 85–106.
4. MINISTRY OF AGRICULTURE, FISHERIES AND FOOD (1982). *Survey of Arsenic in Food* (HMSO, London).
5. DABEKA, R.W., MCKENZIE, A.D. and LACROIX, G.M.A. (1987). Dietary intakes of lead, cadmium, arsenic and fluoride by Canadian adults: a 24-hour duplicate diet study. *Food Add. Contam.*, 4, 89–102.
6. DICK, G.L., HUGHES, J.T., MITCHELL, J.W. and DAVIDSON, F. (1978). Survey of trace elements and pesticide residues in the New Zealand diet. *NZ J. Sc.*, 21, 57–69.
7. GARTRELL, M.J., CRAUN, J.C., PODREBARAC, D.S. and GUNDER-SON, E.L. (1985). Pesticides, selected elements, and other chemicals in adult total diet samples, October 1978–September 1979. *J. Assoc. Off. Anal. Chem.*, 68, 862–75.
8. FOOD ADDITIVES AND CONTAMINANTS COMMITTEE (1984). *Report on the Review of Arsenic in Food Regulations*. MAFF, FAC/REP/39 (HMSO, London).
9. SLORACH, S., GUSTAFSSON, I.B., JORHEM, L. and MATTSSON, P. (1983). Intake of lead, cadmium and certain other metals via a typical Swedish weekly diet. *Var Foda*, 35, Supplement 1, 1–16.
10. BUCHET, J.P., LAUWERYS, R., VANDEVOORDE, A. and PYCKE, J.M.

(1983). Oral daily intake of cadmium, lead, manganese, copper, chromium, mercury, calcium, zinc and arsenic in Belgium: a duplicate meal study. *Food Chem. Toxicol.*, 21, 19–24.

11. CODEX ALIMENTARIUS COMMISSION (1984). *Contaminants. Joint FAO/WHO Food Standards Program. Codex Alimentarius*, Vol. XVII, 1st edn (WHO, Geneva).

12. BOWEN, H.J.M. (1966). *Trace Elements in Biochemistry* (Academic Press, London).

13. DRINKING WATER STANDARDS (1962). *Public Health Service Pub. No.* 956 (US Govt. Printers, Washington, DC).

14. BORGONO, J.M. and GREIBER, R. (1972). *Trace Substances in Environmental Health V*, ed. HEMPILL, D., 13–24 (Uni. Missouri Press, Columbia, Mo.).

15. BEBBINGTON, G.N., MACKAY, N.J., CHVOJKA, R. *et al.* (1977). Heavy metals, selenium and arsenic in nine species of Australian commercial fish. *Aust. J. Mar. Freshwater Res.*, 28, 277–86.

16. NATIONAL HEALTH AND MEDICAL RESEARCH COUNCIL (1986). *Model Food Legislation. A*12. *Metals and Contaminants in Food* (Australian Government Publishing Service, Canberra, ACT).

17. WORLD HEALTH ORGANISATION EXPERT COMMITTEE (1973). *Trace Elements in Human Nutrition. WHO Tech. Rep. Ser. No.* 532, 50 (WHO, Geneva).

18. MONIER-WILLIAMS, G.W. (1949). *Trace Elements in Food*. (Chapman and Hall, London).

19. WICKSTROM, G. (1982). Arsenic emission from the Novaky power station. *Work Environ. Health*, 9, 2–8.

20. *Observer* newspaper, January 1979 (London).

21. FRIBERG, L. (1978). *Intn. Conf. Heavy Metals in Environ. Toronto, Symp. Proc.*, 1, 21–34 (Uni. Toronto).

22. TSUCHIYA, K. (1977). Arsenic contamination of infant formula. *Environ. Health Perspect.*, 19, 35–42.

23. WOOLSON, E.A. (1975). Arsenic in cotton seed byproducts. *J. Agric. Food Chem.*, 23, 677–83.

24. CODE OF FEDERAL REGULATIONS. Title 21. Section 120, 192/3/5/6. (US Government Printing Service, Washington, DC).

25. LANDER, H., HODGE, P.R. and CRISP, C.S. (1965). Arsenic levels in hair. *J. Forensic Med.*, 12, 52–67.

26. INTERNATIONAL AGENCY FOR RESEARCH ON CANCER (1973). *Evaluation of Carcinogenic Risk of Chemicals to Man. Some Inorganic and Organometallic Compounds*, Vol. 2 (IARC, Lyon).

27. RHIAN, M. and MOXON, A.L. (1943). Interaction of arsenic and selenium. *J. Pharmacol. Exp. Ther.*, 78, 249–64.

28. DABEKA, R.W. and LACROIX, G.M.A. (1985). Graphite-furnace atomic absorption spectrometric determination of arsenic in foods after dry-ashing and coprecipitation of arsenic with ammonium pyrrolidine dithiocarbamate. *Can. J. Spectroscopy*, 30, 154–7.

29. NHAM, T.T. and BRODIE, K. (1989). Determination of arsenic and selenium in soil by graphite furnace and vapour generation atomic absorption

spectrophotometry. *Chemistry International*: *International Conference of the Roy. Aust. Chem. Inst.*, 28 August–2 September, Brisbane, *Abstracts* 94.
30. TSALEV, D.S. (1984). *Atomic Absorption Spectrometry in Occupational and Environmental Health Practice*, Vol. II, 15 (CRC Press, Boca Raton, Florida).
31. BROWNING, E. (1969). *Toxicity of Industrial Metals* (Butterworths, London).
32. NIELSEN, F.H. (1986). Other elements: Antimony, in *Trace Elements in Human and Animal Nutrition*, ed. MERTZ, W., 5th edn, Chap. 10 (Academic Press, New York).
33. MONIER-WILLIAMS, G.W. (1934). *Antimony in Enamelled Hollow-ware*, Ministry of Health Report on Public Health and Medical Subjects, No. 7 (HMSO, London).
34. NATIONAL HEALTH AND MEDICAL RESEARCH COUNCIL (1987). *Food Standards Code* (Australian Government Publishing Service, Canberra, ACT).
35. MURTHY, G.K., RHEA, U. and PEELER, J.R. (1971). Antimony in the diet of children. *Environ. Sci. Technol.*, 5, 436–42.
36. SOUKUP, A.V. (1972). Trace elements in water, in *Proc. Cong. Environ. Chem. – Hum. An. Health* (Fort Collins, Colorado, 7–11 August).
37. MACKAY, K.M. (1960). *Hydrogen Compounds of the Metallic Elements* (Spon, London).
38. YANG, G., WANG, S., ZHOU, R. and SUN, S. (1983). Endemic selenium intoxication of humans in China. *Am. J. Clin. Nutr.*, 37, 872–81.
39. KNOTT, S.G. and MCCRAY, C.W.R. (1959). Two naturally occurring outbreaks of selenosis in Queensland. *Aust. Vet. J.*, 35, 161–5.
40. WORLD HEALTH ORGANISATION (1987). *Environmental Health Criteria 58: Selenium.* United Nations Environment Programme, the International Labour Organisation and the World Health Organisation (WHO, Geneva).
41. YANG, G., ZHOU, R., YIN, L. *et al.* (1989). Studies of safe maximal daily dietary selenium intake in a seleniferous area in China. I. Selenium intake and tissue selenium levels of the inhabitants. *J. Trace Elem. Electrolytes Health Dis.*, 3, 77–87.
42. THOMSON, C.D. and ROBINSON, M.F. (1980). Selenium in human health and disease with emphasis on those aspects peculiar to New Zealand. *Am. J. Clin. Nutr.*, 33, 303–23.
43. ROBINSON, M.F., REA, H.M., FRIEND, G.M. *et al.* (1978). On supplementing the selenium intake of New Zealanders. *Br. J. Nutr.*, 39, 589–600.
44. LI, C.Z., HUANG, J.R. and LI, C.X. (1987). Sodium selenite as a preventive measure for Kashin-Beck disease as evaluated in X-ray studies, in *Selenium in Biology and Medicine*, 3rd Symposium, Part B., eds COMBS, G.F., SPALLHOLZ, J.E., LEVANDER, O.A. and OLDFIELD, J.E., 934–7 (Van Nostrand Reinhold, New York).
45. ALFTHAN, G. (1988). Longitudinal studies on the selenium status of healthy adults in Finland during 1975–1984. *Nutr. Res.*, 8, 467–76.
46. LOMBECK, I. (1983). Evaluation of selenium status in children. *J. Inher. Metab. Dis.*, 6, *Suppl.* I, 83–4.
47. GROPPER, S.S., ACOSTA, P. B., CLARKE-SHEENAN, N. *et al.* (1988).

Trace element status of children with phenylketonuria. *J. Am. Diet. Assoc.*, 88, 459–65.

48. GREAVES, C., PATTERSON, C., TINGGI, U. and REILLY, C. (1989). Dietry selenium intake of young Brisbane children. Proc. Nutr. Soc. Australia, 14, 127.

49. THORN, J., ROBERTSON, J., BUSS, D.H. and BUNTON, N.G. (1976). Trace nutrients. Selenium in British food. *Br. J. Nutr.*, 39, 391–6.

50. REILLY, C., PATTERSON, C.M., TINGGI, U. *et al.* (1990). Trace element nutritional status and dietary intake of children with phenylketonuria. *Am. J. Clin. Nutr.*, 52, 159–165.

51. ARTHUR, D. (1972). Metals in Canadian seafish. *Can. Inst. Fd. Sci. Tech.*, 5, 165–8.

52. EGAAS, E. and BRAEKKAN, O.R. (1977). *Fiskeridirektoratets Skrifter Serie Ernaering*, 1 (3), 87–91.

53. BEBBINGTON, G.N., MACKAY, N.J., CHVOJKA, R. *et al.* (1977). Heavy metals, selenium and arsenic in nine species of Australian commercial fish. *Aust. J. Mar. Freshwater Res.*, 28, 277–86.

54. FOOD AND NUTRITION BOARD (1980). *Recommended Dietary Allowances*, 9th edn (National Acad. Sci., Washington, DC).

55. DREOSTI, I.E. (1986). Selenium. *J. Food Nutr.*, 43, 60–78.

56. YOUNG, V.R., NAHAPETIAN, A. and JANGHORBANI, M. (1982). Selenium bioavailability with reference to human nutrition. *Am. J. Clin. Nutr.*, 35, 1076–88.

57. COMBS, G.F. and COMBS, S.B. (1984). The nutritional biochemistry of selenium, in *Ann. Rev. Nutr.*, eds. DARBY, W.J., BROQUIST, H.P. and OLSON, R.E., 257–80 (Ann. Revs. Inc., Palo Alto, Cal.).

58. TAPPEL, A.L. (1980). Vitamin E and selenium protection from *in vivo* lipid peroxidation, in *Micronutrient Interactions: Vitamins, Minerals and Hazardous Elements*, ed. LEVANDER, O.A. and CHENG, L., 18–31 (New York Acad. Sc., New York).

59. WHANGER, P.D. (1985). Metabolic interaction of selenium with cadmium, mercury and silver, in *Advances in Nutritional Research*, ed. DRAPER, H., 221–50 (Plenum Press, New York).

60. SMITH, M.L. and WESTFALL, B.B. (1937). Further field studies on the selenium problem in relation to public health. *US Pub. Health Rep.*, 52, 1375–6.

61. JAFFE, W.G. (1976). Effects of selenium intake in humans and rats, in *Proc. Symp. Selenium–Tellurium in the Environment*, Uni. Notre Dame Industrial Health Foundation (Notre Dame, Ind.).

62. CENTRES FOR DISEASE CONTROL (1984). Selenium intoxication. *Morb. Mort. Wkly. Rep.*, 33, 157–8.

63. LO, M.T. and SANDI, E. (1980). Selenium: occurrence in foods and its toxicological significance — a review. *J. Environ. Pathol. Toxicol.*, 4, 193–218.

64. NATIONAL HEALTH AND MEDICAL RESEARCH COUNCIL (1987). *Food Standards Code*, 71 (Australian Government Printing Service, Canberra, ACT).

65. SCHRAUZER, G.N. and WHITE, D.A. (1978). Selenium and human nutrition: dietary intakes and effects of supplementation. *Bioinorg. Chem.*, 8, 303–18.

66. OLDFIELD, J.E. (1978). Nutrient deficiencies in animals: selenium, in *CRC Handbook Series in Nutrition and Food Science, Section E: Nutritional Disorders, Vol. II, Effects of Nutrient Deficiencies in Animals*, 335–40 (CRC Press, Boca Raton, Florida).

67. CHEN, X., YANG, G., CHEN, X. *et al.* (1980). Studies on the relations of selenium to Keshan disease. *Biol. Trace Element Res.*, 2, 91–100.

68. BURK, R.F. (1984). Selenium, in *Nutrition Reviews: Present Knowledge in Nutrition*, 5th edn, 519–27 (Nut. Found., Washington, DC).

69. VAN RIJ, A.M., THOMPSON, C.D., MCKENZIE, J.M. and ROBINSON, M.F. (1979). Selenium deficiency in total parenteral nutrition. *Am. J. Clin. Nutr.*, 32, 2076–85.

70. SCHRAUZER, G.N., WHITE, D.A. and SCHNEIDER, C.J. (1977). Cancer mortality correlation studies. III. Statistical associations with dietary selenium intakes. *Bioinorg. Chem.*, 7, 23–31.

71. MCORIST, G.D., FARDY, J.J. and FLORENCE, T.M. (1987). Rapid determination of selenium in human serum by neutron activation analysis. *J. Radioanal. Nucl. Chem., Letters*, 119, 449–55.

72. CHENG, Y.D., ZHUANG, G.S., TAN, M.G *et al.* (1988). Preliminary study of correlation of Se content in human hair and tissues. *J. Trace Elem. Expt. Med.*, 1, 19–22.

CHAPTER 8

THE PACKAGING METALS: ALUMINIUM AND TIN

For many years tin and aluminium have been accepted as safe and highly satisfactory metals out of which to make cooking utensils and food containers. Tin has been used in this way, both on its own and, in far greater quantity, as part of the alloy pewter, since Roman times. Though much more recently discovered, aluminium has also contributed to our culinary requirements in a similar way.

There have always been some doubts about whether the two metals were really as 'user friendly', to use the modern term of the computer age, as the community had been led to believe. People talked of 'tin poisoning', especially after canning of food began to be used on a large scale. Though the poisoning was, in fact, often due to the lead solder used to seal the cans and not to tin, the term continued to be used. Aluminium has been subjected to several major campaigns by those who doubted its suitability for use in food preparation and wanted it banned from the kitchen where it had taken over from traditional iron and copper saucepans. But both metals weathered the storms and their use in food processing and preservation kept on increasing.

Now, in our very consumer-sensitive world, where questions of pollution and food hygiene, stimulated by news media hungry for sensation, can bring down ministers of state and threaten governments, aluminium and, to a lesser extent, tin are once more being subjected to searching probes about their safety for use with foods. New evidence is being produced, often as a result of improved analytical techniques, that, perhaps, there is need for a re-assessment of the situation and that stricter control needs to be exercised in the interests of public health.

ALUMINIUM

As was noted by Alfrey[1] in the fifth edition of Underwood's classical text *Trace Elements in Human and Animal Nutrition*[2], only a mere three pages were devoted to aluminium in the fourth, 1977 edition, whereas in 1986 the metal had been given a whole chapter to itself. This was not surprising, since a decade earlier much less had been known about the metabolism, toxicity and distribution of the element in biological tissues and foods. That great measure of scientific interest in a topic, citation in the *Index Medicus*, showed that in the ten years from 1966 to 1976 there were only 80 citings on aluminium, compared to more than 268 in the seven years from 1977 to 1984. The annual citation rate today is greater still as interest in the element continues to grow. As we shall see, there are good reasons for this high level of interest in aluminium among food and health scientists, as well as the general community.

Aluminium is the most common metal in the earth's crust. Because of its reactive nature, it does not occur in a free, elemental state, but in combination with oxygen, silicon, fluoride and other elements as silicates and other compounds. Many of its naturally-occurring compounds are insoluble and thus concentrations of the element in water, both fresh and sea, is normally low, at less than 1 μg/litre. However, acid conditions can increase the solubility of the element. A consequence of 'acid rain' in the northern hemisphere is that aluminium levels in some lakes and rivers have been increasing.

Aluminium was discovered, and named, by Humphrey Davy in 1807. He gave the element its -ium ending, in keeping with the names of two other elements he had discovered, sodium and potassium. The alternative spelling and pronunciation, aluminum, is now widely used, especially by American scientists.

It was not until 1825, that elemental aluminium was actually produced by Oersted in Denmark. Even after that, in spite of the natural abundance of the metal, it required the availability of abundant and cheap electrical energy to allow extraction in commercial quantities from its highly intractable ores. This was first done by the French scientist Deville in 1855. Aluminium's first major public appearance was at the Paris Exhibition of that year where the new, light, silvery metal created a major stir. An extraction method less expensive than Deville's was developed by the German Bayer in 1889. The *Bayer process* is what is used today to produce aluminium from its ores.

In spite of the enthusiastic welcome given to it at the Paris Exhibition, it was some time before aluminium was accepted as a useful addition to industry. It was not until later in the century, when progress in metallurgical techniques allowed casting, rolling and joining of the metal to be carried out efficiently and economically, that this occurred. Once that was achieved, its uses were quickly recognised. Its resistance to corrosion, light weight combined with strength, and its lack of toxicity, made it ideal for constructing cooking and food storage utensils and, later, cans. It was well into the present century before the latter use developed appreciably. Today, it is one of the major uses of the metal.

The growth rate of use of aluminium in recent years has been greater than for any other metal. Primary production worldwide was estimated to have increased from about 4.5 million tonnes in 1960 to nearly 15.5 million tonnes in 1980[3]. Primary production actually decreased in the following five years, to about 14 million tonnes, reflecting a recession in the metals industry and, also, very successful aluminium recycling campaigns. Today recycling of aluminium scrap is a major source of the metal.

CHEMICAL AND PHYSICAL PROPERTIES

Aluminium has an atomic weight of 27 and is number 13 in the Periodic Table. It is a light metal with a density of 2.7. Its melting point is 660.4° C. It is a soft, ductile, silver-white metal, with good electric and heat conductivity. It is extremely resistant to corrosion, though its alloys are less so. It has an oxidation state of 3. Aluminium is a very reactive metal and can ignite in air if mixed with the oxide of other metals, for example of iron, and exposed to heat. This is the basis of the thermite reaction, used in incendiary bombs, in which the heat is supplied by ignition of magnesium ribbon.

The principal inorganic compounds of aluminium are the oxide, hydroxide, sulphate, fluoride and chloride. The metal also forms organic compounds, some of which are highly reactive and become hot and fume on exposure to air.

PRODUCTION AND USES

Though aluminium compounds make up about 7 per cent of the earth's crust, mainly as clays in soils and silicates in rocks, it is commercially extracted only from a few types of ore. These are the bauxites (hydrated aluminium oxide) and cryolite (sodium aluminium fluoride). Major deposits of these are found in the Caribbean region, the Soviet Union, South Africa and Australia.

To extract the metal from bauxite, the ore is converted to aluminium hydroxide by treatment with caustic soda. This is then converted to alumina (aluminium oxide, Al_2O_3) by calcination at about 150° C. The alumina is then mixed with fused cryolite and fluorspar (CaF_2) at high temperatures and subjected to electrolysis. Aluminium is released at the cathode and is run off in a molten state from time to time. This stage of the extraction requires considerable heat and electrical energy.

Aluminium produced in this way is usually cast into solid ingots. These can be reheated and rolled and pressed into a variety of forms, as plate, sheet, wire, and foil for commercial use. In many cases the pure aluminium is combined, while still molten, with other metals to form alloys. The metals used are copper, zinc, chromium, magnesium, nickel, titanium, iron and silicon. Alloying increases the strength and improves other qualities of aluminium, especially as regards casting and machining.

Because of its lightness, electrical conductivity and corrosion resistance, as well as other properties, aluminium is the most widely used of all the primary metals. About half the world's production is used in the electrical industry. Much is also used in automobile engineering, aircraft and ship building, building construction, and in the production of domestic appliances and other goods. These range from simple cooking and storage utensils, foil and take-away food containers, to cookers, refrigerators, air-conditioners and many other such modern household appliances. Another major non-construction use of aluminium is in cans, especially for beverages.

Aluminium compounds have many uses[4]. The food industry uses them as food additives, in baking powder, processed cheese, manufactured meats and other products. Aluminium is widely used in the pharmaceutical and cosmetics industries, in toothpaste, anti-perspirants, a variety of therapeutic agents and related products. Aluminium sulphate and other compounds are used for particle sedimentation in water treatment. Other uses are in the production of high temperature-resistant refractories, ceramics and whiteware.

ALUMINIUM IN FOOD AND BEVERAGES

Until relatively recently little reliable information was available on levels of aluminium in foods and diets. As Greger cautioned[5], earlier data on the aluminium content of foods should be viewed with scepticism because of the possibility of over-estimates due to faulty analytical methodology. In support of this view she drew attention to the fact that estimates of aluminium in blood plasma had decreased by more than 50-fold in the previous 20 years because of improvements in analytical methodology.

Three compilations of published data on the aluminium content of foods and diets were published over the past three decades. One by Campbell and colleagues[6] included analyses published between 1929 and 1955; a second, by Sorenson and colleagues[7], covered publications from 1932 to 1973 and, the third, by Schlettwein-Gsell and Mommsen-Straub[8], updated the first two reviews.

A recently published extensive review by Pennington[9], has given us a further valuable update on the situation and clarified many points about levels and sources of aluminium in foods and diets.

Of particular interest in Pennington's review is the conclusion that, whereas in pre-1980 reports, daily intakes of aluminium in the US were estimated to be 18–36 mg, more recent and probably more accurate data indicate a lower level of intake of 9 mg/day for teenage and adult females and 12–14 mg/day for teenage and adult males.

Pennington's extensive data cover mainly results reported from the US and Finland from 1932 to the time of writing. Of particular importance is information on aluminium concentrations in 234 foods sampled in the Food and Drug Administration's 1984 Total Diet Study (TDS)[10].

Selected data from the US TDS study given in Pennington's review are (means expressed as mg/kg):

—Cow's milk: 0.06; cheese: cheddar 0.19; American (processed): 411.0; infant formula: milk based 0.05;
—Meat: beef 0.28; meatloaf: 1.26; bacon: 3.63; sausage: 1.82; chicken: roast 0.22;
—Fish: cod 0.47; fish sticks: 51.4; tuna: canned, 0.67;
—Fruit: apples 0.14; cherries: 0.19; grapes: 1.81;
—Vegetables: broccoli 0.98; beets: canned 0.26; carrot: 0.19;
—Cereals: corn grits 0.2; oatmeal: 0.68; rice: 1.42;
—Cereal products: biscuits 16.3; bread: white 2.33; bread: rye 4.07; pancakes: from mix 69.0; tortilla: flour 129.0; cake: chocolate 86.0;
—Other: milk chocolate 6.84; chocolate chip cookies: 5.61;
—Tea: from bag 4.46;
—Water: domestic, eastern US 0.12.

Pennington's own data, as well as the results from many other studies which she summarises, indicate that generally levels of aluminium in foods are low. However, there are a few food items, including certain spices as well as tea, which naturally accumulate higher than average levels of the element. As the results show, the tea plant, *Camellia sinensis*, is one of these natural accumulators of aluminium.

Numerous aluminium compounds are used as food additives. Sodium aluminium phosphates are the most widely used of these substances. They are added to baking powder as acidifying agents. In processed cheese they are used as emulsifiers and in meat as binders. Aluminium silicates are used as anticaking agents in salt, non-dairy creamers and other dry products. Other uses of aluminium compounds are as stabilisers, thickeners, bleaching agents and texturisers. Thus many manufactured food products can contain far more aluminium than their original food components would indicate. Pennington has noted that 'additives seem to have the most dramatic effects in increasing aluminium levels. The major food sources of aluminium in daily diets are probably grain products with aluminium additives, processed cheese with aluminium additives, tea, herbs, spices and salt'.

Water is not normally an important source of dietary aluminium. As has been noted above, aluminium compounds are generally only slightly soluble in water, though solubility can increase under acid conditions. However, alum (aluminium sulphate salts) are widely used as a coagulating agent in domestic water supplies. It has been shown that the use of alum at normal accepted levels, can increase aluminium levels three-fold, from about 0.2 mg/litre to 0.6 mg/litre[11]. If, through accident, addition of alum to water is greatly increased, as has occurred in recent years in the UK, or where the water is acidic, levels of aluminium in the water can be appreciably increased. The use of aluminium-containing domestic water in kidney dialysis machines and in total parenteral nutrition fluids and other pharmaceutical preparations, has been found to be a source of increased intake of the metal in some people[12].

Studies by several investigators[13] have shown that the use of aluminium cooking utensils, as well as of foil, can increase the amount of aluminium in some, but not all, foods. Uptake depends to a large extent on the acidity of the food. Potatoes, for instance, showed no increase when boiled in an aluminium saucepan, compared to a glass pan. Stewed tomatoes, however, with 0.14 mg/kg after cooking in glass, had 15.5 mg/kg after use of an aluminium saucepan. Cabbage, cooked in similar ways, increased from 0.34 to 90.8 mg/kg. Increases in aluminium in meat stored and cooked in different utensils were not so significant, but still occurred. Raw, frozen beef contained 0.39 mg/kg of aluminium; when frozen in aluminium foil, its content went up to 0.47; when cooked in the foil, levels rose to 5.7, and when reheated in the foil, again increased to 7.4 mg/kg.

Beverages can also take up aluminium from containers. Beer in a tin plate can contained 0.15 mg/kg of aluminium; a similar beer in a tin plate can with aluminium ends had 0.38 mg/kg. Fresh orange juice contained 1.3 mg/kg of the metal; juice in a wrought aluminium can 12.4 mg/kg and in a cast aluminium can 47.4 mg/kg. Similar increases were found in peach, tomato and several other low pH juices. Aluminium concentrations in water boiled in an aluminium pan, at pH 3, increased in 20 min from 0.05 to 8.08 mg/kg[14].

As Pennington has observed, while migration of aluminium from cooking vessels and foil does occur, this route of dietary uptake is probably neither consistent nor of major significance. However, under certain circumstances, as will be seen, even an apparently insignificant increase of aluminium in food or water could have serious implications for people with certain health problems.

Not everyone will be satisfied with Pennington's reassurance regarding the lack of significance for human health of aluminium uptake from cooking utensils. The opposition to use of aluminium in this way has been continuing for a long time and has recently gathered momentum once again. As early as 1886, only shortly after the metal had begun to be used extensively for making cooking utensils, claims were made that ingestion of aluminium resulted in toxic effects. Considerable controversy resulted and continued for many years[15]. Forty years later the controversy still continued and was still unresolved[16]. In the following year Spira reopened the attack on aluminium with evidence, he believed, that substantiated earlier criticisms[17]. In 1971 the American Medical Association felt obliged to enter the field with a statement that evidence was lacking to support the view that the small amounts of aluminium ingested in the diet, even when aluminium cooking utensils were used, had any ill effects on human health[18]. This view was substantially repeated by the US Food and Drug Administration some years later[19].

The controversy refuses to go away. As will be seen, reports in the medical literature, as well as in the popular media, of connections between aluminium and a variety of diseases, as well as of high intakes of the metal in water in parts of the UK and elsewhere, makes it an issue of considerable interest to the community. While there would seem to be no reason to doubt Pennington's reassurance regarding the general safety of aluminium in cooking and other catering utensils, there are good grounds for believing that the question must be kept under review and, if necessary, stringent regulations introduced regarding leaching of aluminium into food.

ABSORPTION AND METABOLISM OF ALUMINIUM

In spite of the widespread distribution of aluminium in the environment and in foods, little is absorbed or retained by the body. Why this is so, is not well understood. The chemical form of the element is an important factor in controlling absorption. Aluminium phosphate is very insoluble and is poorly absorbed, as is the hydroxide. Aluminium citrate, on the other hand, is readily absorbed from the gut. Both vitamin D and the parathyroid hormone play a part in moderating aluminium absorption. Body stores of iron also appear to be important. In patients with low ferritin levels, there is an increased absorption of aluminium.

Gormican and Catli[20] found that 7 per cent of ingested aluminium was absorbed by healthy young men consuming a diet containing 10.33 mg of the element over a 28-day period. Aluminium balances (net body retention) over the period was minimal. Greger and Baier[21] found similar results in another study.

The gastro-intestinal tract is a major barrier to aluminium absorption. Where this is by-passed, as in total parenteral nutrition, a high body burden of the metal can result. This has also occurred in uremic patients on dialysis where aluminium-contaminated domestic water was used in the dialysate. Aluminium in the dialysate readily transferred across the dialysing membrane into the patient, with consequences which will be discussed below[22]. The kidneys normally are able to excrete absorbed aluminium efficiently, but in kidney failure, this does not occur. Chronic, renal failure patients are normally prescribed aluminium hydroxide-containing gels to limit absorption of phosphate from the gastro-intestinal tract. This is done to prevent the build up of high levels of serum phosphate which is a symptom of renal failure. Thus, these patients, with inadequate kidney function and a high intake of aluminium, can accumulate a large body burden of the metal.

It has been shown that when individuals are given large oral loads of aluminium, some of the excess is absorbed by the body[23]. Urinary output increases, but some storage also takes place. Tissues most affected are liver, kidney, spleen and bone. Some may also be stored in the brain and heart.

Aluminium toxicity is a subject of considerable debate at the present time. It is known to be toxic, under certain conditions, to plants and fish. There is evidence that it can also be toxic to humans. The literature on the subject is extensive, but it has been briefly summarised by Pennington[10], to whom the following paragraph owes a great deal.

Accumulation of aluminium in the brain has been associated with Alzheimer's disease[24]. Dialysis dementia has been reported to result from use of aluminium-containing water in kidney dialysis and from ingestion of aluminium-containing phosphate-binding gels by renal patients[25]. High intakes of aluminium have also been associated with osteomalacia and bone fractures[26]. Other aluminium-disease associations listed by Pennington include metabolic alkalosis, Parkinsonism-dementia, and bowel obstruction. These illnesses have been associated with large aluminium intakes resulting from ingestion of certain pharmaceutical products, such as antacids and phosphate-binding gels, from use of aluminium-contaminated fluids in dialysis and parenteral nutrition, as well as with impaired aluminium excretion in renal disease.

Whether illness similar to those described above, or any other health risk, can arise from dietary intake by healthy individuals of the amounts of aluminium normally found in foods, has not been established. What is clear is that there are certain people who have chronic kidney disease and take large doses of aluminium-containing drugs, as well as others maintained on total parenteral nutrition, who are at risk of developing aluminium intoxication and whose intake of aluminium-containing foods should be controlled. There is little evidence that the majority of people need to take such precautions.

ANALYSIS OF FOODSTUFFS FOR ALUMINIUM

As has been mentioned, there are considerable technical difficulties in analysing foods for aluminium, especially at low levels of concentrations. Some of these have been overcome by the development of better analytical equipment and techniques, but two major difficulties remain: how to overcome the very real problem of contamination from all-intrusive aluminium in the environment, and the absence of a range of reliable, certified reference materials. Savory and Wills[27] have discussed these problems.

Sample preparation for analysis may be carried out by wet digestion. Preconcentration by solvent extraction is recommended for low concentrations of the element. Dry ashing causes loss of aluminium chlorides. Microwave digestion is recommended by Savory and Wills as the most promising advance in sample preparation. They use a commercial, microwave-powered, teflon-lined, sealed bomb. The bomb eliminates contamination and volatilisation losses.

Graphite furnace AAS has been effectively applied to aluminium determination in biological fluids and foods. Levels in blood serum have been accurately determined with GFAAS by Gardiner and his colleagues[28]. In the author's laboratory aluminium in milk, other dairy products and soya products have been analysed by Zeeman mode background corrected AAS[29].

Foods in the Food and Drug Administration's TDS, referred to above, were analysed using inductively coupled plasma. Where it is available, ICP is undoubtedly a method of choice for aluminium determination. It overcomes matrix problems, due to the refractive nature of aluminium, which can cause some difficulties when using GFAAS.

Neutron activation analysis, again where instrumentation is available, can be used effectively to determine aluminium in foods. Extensive sample preparation is not necessary with some samples, which is a considerable advantage and reduces the risk of contamination. However, in samples with a high phosphorus content, a preliminary separation step is necessary to remove this element.

TIN

Tin is one of the ancient metals. Long before the Romans crossed from Gaul to Britain, Phoenician merchant ships had been calling at Cornish ports to load the rare and valuable metal that the Celtic miners had won from their galleries and pits. The metal was valued because of its ability to form alloys with distinctive and very useful properties when mixed with other metals. Soft and vulnerable copper was converted into tough and resistant bronze by the addition of tin; lead was converted into rigid, and lighter pewter. These were the sort of metallurgical innovations which helped the ancient empires win their wars and expand their commercial powers.

It was only in relatively recent times that tin has been used extensively in its own right. For the last 150 years considerable quantities of pure tin have been used, to plate containers and other utensils used to store and process food.

Even before tinplate was developed, the metal had been used extensively in the kitchen and on the dining table. Bronze was used for making cooking and storage vessels, as well as eating implements for many hundreds of years. Pewter plates were on the tables of rich man and commoner since the alloy was first discovered right up to the introduction of porcelain and glazed earthenware on an extensive scale in the seventeenth century. Thus, man has been eating food which has been in contact with tin, and which had probably absorbed some of the metal to add to what it already naturally contained, for many hundreds of years. This culinary use of tin has stood the test of time well, from practical as well as health points of view. Only in some exceptional cases has the presence of tin in food resulted in serious poisoning. Even today, in spite of some recent expressions of concern, it is still generally accepted as safe for use with food. Indeed, there is evidence that tin may be an essential trace nutrient in some animals, possibly including man.

CHEMICAL AND PHYSICAL PROPERTIES

Tin has the symbol Sn after its Latin name *stannum*. It is element number 50 in the Periodic Table, with an atomic weight of 118.7. Tin is normally a mixture of three crystalline forms and, consequently, its specific gravity is not fixed but lies between 5.8 and 7.3, depending on the proportion of each form present. It is a soft, white, lustrous metal which is easily rolled into foil and extruded into tubes. Its melting point is low at 231.9° C. Tin is highly resistant to corrosion.

Oxidation states of tin are 2 and 4. It forms two series of compounds, stannous and stannic. Among its more important inorganic compounds are the two chlorides, tin oxide (stannous oxide, SnO_2), sodium chlorostannate, sodium pentachlorostannate, stannous fluoride and sodium pentafluorostannate.

Tin also forms a number of organic compounds, the organotins. These are important industrially, especially the alkyl and phenyl compounds.

PRODUCTION AND USES

Though tin is widely distributed in small amounts in most soils, it is commercially produced in quantity in only a few places in the world. Its principal ore is casserite or tinstone, the dioxide, SnO_2. Tin also occurs in some other ores, in combination with, for example, tungsten. Extraction of tin from mixed ores is difficult, but is undertaken where the quantity and value of the other metals are sufficient to justify the high processing cost.

Tin is produced on a large scale in Malaysia, Bolivia, Thailand, Indonesia and the Soviet Union. Small-scale production is undertaken in several other countries, but this is usually intermittent and depends on world prices for tin. The ancient Cornish mines have not been in commercial use for some years.

Upwards of 50 per cent of the world's production of tin, which is about 200000 tonnes per annum, is used for plating steel and other metals. Today tinplate is normally manufactured by electrodeposition. This technique has replaced the old method of hot dipping. Tin is resistant to corrosion by food and thus is used to cover the insides of food and beverage containers and some cooking and food processing equipment. Tin coatings are also used in the manufacture of engineering and electrical components and for other industrial applications where corrosion resistance is important. Alloys of tin with other metals, for example bronzes of various kinds, are used in a variety of industries. Pewter, which contains about 90 per cent tin, and various related alloys of tin with antimony, copper, lead and some other metals, such as Monel and Britannia metal, are used largely for ornamental and some catering purposes.

About 5 per cent of total tin production is used in the chemical industry. It is used in the manufacture of certain types of glass and enamels, in dyeing and printing of textiles, and as a reducing agent in certain chemical processes. The pharmaceutical industry also uses some tin compounds in such products as toothpaste. Organic compounds of tin are used as stabilisers in polyvinylchloride plastics, as well as in rubber paints. Other organic compounds are used as fungicides and other purposes in agriculture. An important, and growing use, which will be discussed later because of its health implications, is of organotin compounds as anti-fouling agents on ships.

TIN IN FOOD AND BEVERAGES

The normal level of tin in foods is low, except where they have been in contact with the metal in cans or other tin-plated containers. Results of the 1982 UK Total Diet Study, as reported by Sherlock and Smart[30], show that means levels of tin in all major food groups were less than 1.0 mg/kg, with the majority <0.2 (range <0.2–13), except for canned vegetables (mean 32, range 9–80 mg/kg) and fruit products (mean 40, range 12–129 mg/kg).

Sherlock and Smart point out that tinplate, as used in food containers, is, in fact, only slowly attacked by food with which it is in contact, unless certain substances are present which accelerate corrosion. These are cathode depolarisers, such as residual and nitrate oxygen, and their presence can increase the rate of breakdown considerably, resulting in can contents with unacceptably high levels of tin[31]. Acid conditions can also contribute to corrosion[32]. Coating the inside of cans with lacquer will reduce the likelihood of this corrosion occurring.

Uptake of tin by foods depends on the nature of the foods as well as on whether cans are lacquered or not. Meat and meat products are often contained in lacquered cans and are not aggressive to tin. Relatively low average levels, of about 20 mg/kg, were found in a study of canned food sold in ordinary retail outlets in the UK[31]. In contrast, tomatoes sometimes contain high levels of nitrate and are often sold in unlacquered cans. These products can contain more than 50 mg/kg of tin, and sometimes over 100 mg/kg. Pale fruits, such as grapefruit and pineapple, are also contained in unlacquered cans and are moderately aggressive towards tin. About 20 per cent of samples analysed in the study had levels of over 100 mg/kg with a mean of 75 mg/kg. Highly coloured fruit, such as blackcurrants and raspberries, contain anthocyanins which are aggressive cathode depolarisers. For this reason they are placed in lacquered cans but, nevertheless, were found to have an average tin concentration of about 40 mg/kg, with 10 per cent over 100 mg/kg.

Estimates of tin intake, based on the UK Total Diet Survey, have also been given by Sherlock and Smart. Estimated total weight of food consumed by an adult was 1.435 kg/day, made up of different amounts of foods from the groups covered in the UK TDS[33]. Concentrations of tin in each of these food groups as consumed, as well as in the overall diet, were estimated. Two methods of calculating intakes were used. When a concentration was less than the limit of detection, this was taken to mean either that the food contained tin at the limit, or, at zero concentration. Both estimates were given, as an upper and a lower bound, with 'true' intake lying between the two extremes. Thus, overall intake was estimated to be, upper bound 3.08, lower bound 2.92 mg/day. These results are close to the 3.5 mg/day reported in the US by Schroeder and colleagues[34], but less than the 15 mg/day reported for New Zealand[35].

It is of significance that, while the results of the UK TDS show that canned vegetables and fruit make up only 5 per cent by weight of total food intake, these foods contribute 85 per cent of the total intake of tin. This finding underlies the fact that, in the absence of canned foods, intake of tin is very low. A PTDI of 2 mg/kg body weight has been established by the WHO for tin[36]. The estimated UK intake of about 3 mg/day would, for a 70-kg man, work out at 14 mg of tin each day, more than three times the estimated intake. This appears to confirm the validity of the assurance of the UK's Food Additives and Contaminants Committee that there is no reason to be concerned about current levels of tin in the diet, or in canned foods[37]. However, since individual cans have been found to contain in excess of 250 mg/kg of tin, the maximum permitted level in foods in many countries, the Committee has, in fact, recommended that the permitted level be reduced to 200 mg/kg in tinned food.

There is some concern that while intake of inorganlc tin may be at acceptable levels, this may not always be the case with organotin compounds. Sherlock and Smart[30] believe that it is most unlikely that tin in canned food is anything other than inorganic tin salts. However, there is evidence that certain foods may also contain organic compounds of the element. This does not appear to result from natural conversion of the inorganic to organic forms and accumulation in the food chain, as is the case with organomercury compounds in marine organisms. It results from the increasing industrial use of organotin compounds. Dibutyltin salts are used as stabilisers in PVC plastics and trialkyl and triaryltin compounds as biocides. Tributyltin (TBT) is used extensively as a marine antifouling agent. Tributyltin-based antifouling paints are used on more than 70 per cent of the world's deep-ocean ships as well as on many smaller recreational vessels. As a consequence many boating harbours are highly contaminated with the compound. An Australian study[38] has shown that the waters of Sydney Harbour and the Georges River estuary are significantly contaminated with TBT. Oysters from the commercial beds of the Georges River contain up to 500 times the level of TBT found in oysters from unpolluted waters. Concentrations, expressed as ng Sn/g, ranged from 80 to 130 ng in the contaminated oysters, compared to below 2 ng in those living in unpolluted water. A UK study[39] showed that similar problems exist in the English River Crouch. Though levels of TBT in oysters are in fact low, and consumption of these molluscs limited, health authorities are keeping an eye on the problem. In Australia the use of TBT-based antifouling paints on certain types of boats is now banned in several states, mainly to protect the interests of oyster farmers. The health of human oyster eaters may also benefit from the move.

ABSORPTION AND METABOLISM OF TIN

Tin in food appears to be poorly absorbed and is excreted mainly in faeces[40]. At most about 1 per cent of ingested tin enters the blood. Absorption appears to be related to the chemical form of the element, with Sn(II) being four times more readily absorbed than Sn(IV) compounds. Absorbed tin is rapidly excreted initially, but small amounts may be retained in kidney, liver and bone, with little in soft tissue. Excretion is mainly in urine with a little in bile.

Tin has not been shown to have any biological role in the body, though it is known to cause significant growth stimulation in rats. It is believed to be an essential nutrient for these animals, with a daily requirement of 1–2 mg/kg body weight[41]. There is no evidence that man has a similar requirement for tin, though its universal distribution in foods and in human tissues suggests that this might be so.

The toxicology of tin has been extensively investigated[42]. The low toxicity of ingested tin, especially in small quantities, has been well established in experimental animals. High levels of tin in food can cause acute poisoning. The toxic dose in humans has been reported to be 5–7 mg/kg body weight. However, there appears to be considerable individual variation in response to doses of tin. Nausea and vomiting have been reported to occur when levels of 250 mg/kg of tin in food have been consumed. However, levels of as high as 500 mg/litre in fruit juice failed to cause gastric upset in a group of volunteers. Mild toxicity has been observed when intake was increased to more than 1 g/litre. An intake of 200 mg/kg in food consumed over an extended period of time failed to cause toxicity in others[43].

Prolonged intake of tin at subacute levels by rats has been shown to be associated with growth retardation, anaemia and histological changes in the liver. Tin is believed to interfere with iron absorption and haemoglobin formation. Dietary tin has been shown to interfere with the metabolism of certain other elements, besides iron. Rats fed 200 mg/kg of tin in their feed lost more zinc in faeces and retained less in their bones than control animals[44]. A high intake of tin exacerbates symptoms of zinc deficiency in rats on a low zinc diet[45].

Because of such evidence, concern has been expressed by some authorities about the legally-permitted level of tin in food of 200 or even 250 mg/kg in some countries. It has been suggested that this limit does not allow an adequate margin of safety[46].

ANALYSIS OF FOODSTUFFS FOR TIN

Difficulties have been experienced in the analysis of tin by many investigators. Modern analytical instruments and procedures have improved the situation, but the element still requires considerable skill if it is to be determined with accuracy.

Graphite furnace AAS, with deuterium lamp or Zeeman background correction, has been shown to be an effective analytical procedure. However, high ashing temperatures are required and matrix interference can be severe with different tin compounds. The use of pyrolytically-coated graphite tubes and *in situ* coating is recommended, allowing lower atomisation temperatures and higher sensitivity[47].

Evans and colleagues[48] describe a hydride-generation AAS technique which has been used extensively in the UK's TDS. This is a sensitive method for low levels of tin. Tin hydride is generated from acidic digests of the food by $NaBH_4$.

REFERENCES

1. ALFREY, A.C. (1986). Aluminum, in *Trace Elements in Human and Animal Nutrition*, 5th edn., ed. MERTZ, W., Vol. 2, Chap. 8, 399–413 (Academic Press, New York).
2. UNDERWOOD, E.J. (1977). *Trace Elements in Human and Animal Nutrition*, 4th edn, 430–3 (Academic Press, New York).
3. BUREAU OF MINES (1987). *Mineral Year Books*, US Dept of the Interior (Govt. Printing Office, Washington, DC).
4. ALUMINUM COMPANY OF AMERICA (1969). *Aluminum* (Alcoa, Pittsburgh).
5. GREGER, J.L. (1985). Aluminum content of the American diet. *Food Technol.*, 39, 73–80.
6. CAMPBELL, I.R., CASS, J.S., BOUGIE, R. and KEHOE, R.A. (1957). Aluminum in the environment of man: a critical review of its hygienic status. *Arch. Indust. Health*, 15, 359–448.
7. SORENSON, J.R.J., CAMPBELL, I.R., TEPPER, L.B. and LINGG, R.D. (1974). Aluminum in the environment and human health. *Environ. Health Perspect.*, 8, 3–95.
8. SCHLETTWEIN-GSELL, D. and MOMMSEN-STRAUB, S. (1973). Spurenelemente in lebensmittel. XII: aluminium. *Intn. Zeitchr. f. Vit. Ernahrungsforsc.*, 43, 251–63.
9. PENNINGTON, J.A.T. (1987). Aluminum content of foods and diets. *Food Add. Contam.*, 5, 161–232.

10. PENNINGTON, J.A.T. and JONES, J.W. (1988). Aluminum in American diets, in *Aluminum in Health, a Critical Review*, ed. GITELMAN, H.J., 67–100 (M. Dekker, New York).
11. MILLER, R.G., KOPFLER, F.C., KELTY, K.C. *et al.* (1984). The occurrence of aluminum in drinking water. *J. Am. Water Works Assoc.*, 76, 84–91.
12. KLEIN, G.L., ALFREY, A.C., MILLER, N.L. *et al.* (1982). Aluminum loading during total parenteral nutrition. *Am. J. Clin. Nutr.*, 35, 1425–9.
13. GREGER, J.L., GOETZ, W. and SULLIVAN, D. (1985). Aluminum levels of foods cooked and stored in aluminum pans, trays, and foil. *J. Food Protect.*, 48, 772–7.
14. SAMSAHL, K. and WESTER, P.O. (1977). Metal contamination of food during preparation and storage: development of methods and some preliminary results. *Sci. Total Environ.*, 8, 165–77.
15. GLAISTER, G. and ALLISON, A. (1913). Aluminium in food. *Lancet*, 1, 843.
16. BURN, J.H. (1932). *Aluminium and Food. A Critical Examination of the Evidence Available as to the Toxicity of Aluminium.* British Non-Ferrous Metals Research Association. Research Reports. External Series No. 162. (BNFMRA, London).
17. SPIRA, L. (1933). *The Clinical Aspects of Chronic Poisoning by Aluminium and its Alloys* (Bale and Danielsson, London).
18. AMERICAN MEDICAL ASSOCIATION COUNCIL ON FOOD AND NUTRITION (1951). Aluminum in food. *J. Am. Med. Assoc.*, 146, 477.
19. FOOD AND DRUG ADMINISTRATION (1971). *Safety of Cooking Utensils, Fact Sheet* (US FDA, Washington, DC).
20. GORMICAN, A. and CATLI, E. (1971). Mineral balance in young men fed a fortified milk-base formula. *Nutr. Metabol.*, 13, 364–77.
21. GREGER, J.L. and BAIER, M.J. (1983). Excretion and retention of low or moderate levels of aluminum by human subjects. *Food Chem. Toxicol.*, 21, 473–7.
22. ALFREY, A.C. (1989). Physiology of aluminum in man, in *Aluminum and Health: a Critical Review*, ed. GITELMAN, H.J., 101–24 (M. Dekker, New York).
23. KAEHENY, W.D., ALFREY, A.C., HOLMAN, R.E. *et al.* (1977). Aluminum absorption. *Kidney Int.*, 12, 361–5.
24. CRAPPER, D.R., KRISHNAN, S.S. and DALTON, A.J. (1973). Brain aluminum distribution in Alzheimer's disease and experimental neurofibrillary degeneration. *Science*, 180, 511–13.
25. PLATTS, M.M., GOODE, G.C. and HISLOP, J.S. (1977). Composition of the domestic water supply and the incidence of fractures and encephalopathy in patients on home dialysis. *Br. Med. J.*, 2, 657–60.
26. BOYCE, B.F., ELDER, H.Y., ELLIO, H.L. *et al.* (1982). Hypercalcaemic osteomalacia due to aluminium toxicity. *Lancet*, ii, 1009–13.
27. SAVORY, J. and WILLS, M.R. (1989). Analytical techniques for the analysis of aluminum, in *Aluminum and Health: a Critical Review*, ed. GITELMAN, H.J., 1–27 (M. Dekker, New York).
28. GARDINER, P.E., OTTAWAY, J.M., FELL, G.S. and HALLS, D.J. (1981). Determination of aluminium in blood plasma or serum by electrothermal atomic absorption spectrometry. *Anal. Chim. Acta.*, 128, 57–66.

29. TINGGI, U., REILLY, C. and PATTERSON, C. (1989). Aluminium in milk and soya products. *Proc. Nutr. Soc. Australia*, 14, 132.
30. SHERLOCK, J.C. and SMART, G.A. (1984). Tin in foods and the diet. *Food Add. Contam.*, 1, 277–82.
31. BRITTON, S.C. (1975). *Tin versus Corrosion*. Intern. Tin. Res. Inst. Publication No. 510 (ITRI, Greenford, Middlesex).
32. MONIER-WILLIAMS, G.W. (1949). *Trace Elements in Food* (Chapman and Hall, London).
33. PEATTIE, M., BUSS, D., LINDSAY, D.G. and SMART, G.A. (1983). Reorganisation of the British total diet study for monitoring food constituents from 1981. *Food Cosmet. Toxicol.*, 21, 503–7.
34. SCHROEDER, H.A., BALASSA, J.J. and TIPTON, I.H. (1964). Abnormal trace elements in man: tin. *J. Chron. Dis.*, 17, 483–502.
35. DICK, G.L., HUGHES, J.T., MITCHELL, J.W. and DAVIDSON, F. (1978). Survey of trace elements and pesticides in the New Zealand diet. *N.Z. J. Sci.*, 21, 57–69.
36. WORLD HEALTH ORGANISATION (1982). *Evaluation of Certain Food Additives and Contaminants*. Tech. Rep. Ser. No. 623 (WHO, Geneva).
37. FOOD ADDITIVES AND CONTAMINANTS COMMITTEE (1983). *Report on the Review of Metals in Canned Foods*. FAC/REP/38 (HMSO, London).
38. BATLEY, G.E., FUHUA, C., BROCKBANK, C.I. and FLEGG, K.J. (1989). Accumulation of tributyltin by the Sydney rock oyster, *Saccostrea commercialis*. *Aust. J. Mar. Freshwater Res.*, 40, 49–54.
39. THAIN, J.E. and WALDOCK, M.J. (1986). The impact of tributyltin (TBT) in antifouling paints on molluscan fisheries. *Mat. Sci. Technol.*, 18, 193–202.
40. WORLD HEALTH ORGANISATION (1973). *Trace Elements in Human Nutrition*, Tech. Rep. Ser. No. 532, 38–9 (WHO, Geneva).
41. SCHWARTZ, K., MILNE, D.B. and VINEYARD, E. (1970). Growth effects of tin compounds in rats maintained in a trace element-controlled environment. *Biochem. Biophys. Res. Comm.*, 40, 22–9.
42. BARNES, J.M. and STONER, H.B. (1959). The toxicology of tin. *Pharmacol. Rev.*, 11, 211–30.
43. DE GROOT, A.P., FERON, V.J. and TIL, H.P. (1973). Toxicology of tin compounds. *Food Cosmet. Toxicol.*, 11, 19–30.
44. GREGER, J.L. and JOHNSON, M.A. (1981). Effect of dietary tin on zinc retention and excretion in rats. *Food Cosmet. Toxicol.*, 19, 163–6.
45. REILLY, C. (1982). Effects of dietary tin on zinc accumulation and enzyme activity in rats. 12 *Intn. Biochem. Cong.*, *Abstracts* (Perth, Australia).
46. DAVIDSON, S., PASSMORE, R., BROCK, J. and TRUSWELL, A.S. (1975). *Human Nutrition and Dietetics*, 5th edn, 140 (Churchill-Livingstone, Edinburgh).
47. TSALEV, D.L. (1983). *Atomic Absorption in Occupational and Environmental Health Practice*. Vol. II, 204–6 (CRC Press, Boca Raton, Florida).
48. EVANS, W.H., JACKSON, F.J. and DELLAR, D. (1979). Evaluation of a method for the determination of total antimony, arsenic and tin in foodstuffs using measurement by atomic absorption spectrophotometry with atomisation in a silica tube using the hydride generation technique. *Analyst*, 104, 16–34.

CHAPTER 9

TRANSITION METALS

Chemists classify metals in terms of electronic structure, either as *representative* or *transition* metals. The former have all their valence electrons in one shell, whereas the transition metals have valence electrons in more than one shell. This structure gives transition metals variability in their oxidation states, a property which makes them of considerable biological significance. Metals with incomplete underlying electron shells are usually coloured, with the colour depending on the oxidation state. The colour can also be modified by the nature of non-metallic elements or groups with which the metal may be combined. These colours are the basis for the many colorimetric methods of determination which have been developed for transition metals.

As we shall see, transition metals are of considerable biological significance. Also, since they have usually been relatively easy to detect, both qualitatively and quantitatively, because of the distinctive coloured compounds they form with reagents, the biological significance of several of them has been recognised for some time and much is known about their functions and properties in foods and diets.

The transition metals[1] are subdivided by chemists into three groups: the main transition or d block elements, the lanthanide elements, and the actinide elements. Here we are principally interested in the main transition group. This contains elements in which the d shells only are partially filled. Scandium, with an outer electron configuration of $4s^2 3d$, is the lightest member of the group. The eight following elements — titanium, vanadium, chromium, manganese, iron, cobalt, nickel and copper — have partly filled $3d$ shells, either in the ground state of the free atom (all except copper) or in one of their more important ions (all except scandium). This group of elements makes up the first transition series, numbers 21–29 in the Periodic Table.

194

The next element, number 30, zinc, is sometimes considered to belong to the transition metals, since it shares some of their properties, but, strictly speaking, it is not. Its outer electron configuration is $3d^{10}4s^2$ and it forms no compounds in which the $3d$ shell is ionised. It also differs from the transition metals in that, with its filled outer shell, it does not form coloured compounds. However, zinc is usually associated with the transition metals in nutrition and food science texts, and that convention will be followed here.

Two other characteristics of transition metals are of particular interest in relation to the present study. The first is their strong metallic quality: they are mainly hard, tough metals, with high melting points and are capable of conducting heat, and electricity, well. They can also form very useful alloys among themselves and with other metals. This has meant that the transition metals have long been used in the catering and food processing industry for construction of equipment.

The second characteristic is of interest from the point of view of chemical and biological activity. Transition metals, as they go from left to right across the Periodic Table, show only small variations in their physical and chemical properties. This is because, generally, succeeding elements in the series differ in electron configuration by one electron in the next to outer valence shell rather than the outer valence shell. The interchangeability of several of the transition elements in their biological roles, for example as coenzymes, may be accounted for by this characteristic.

We will be concerned mainly with metals of the first transition series, and will pay particular attention to those of most significance in relation to food. These metals are copper, iron, chromium, manganese, nickel and cobalt.

We will look in some detail also at molybdenum and silver from the second series. The others, which include some of the rare metals, will be mentioned very briefly, if at all. We have already looked at some of the radioactive metals which occur in the third transition series and will not consider them further.

COPPER

Copper was probably among the first metals to be discovered and used by humans. This was because it is relatively easily extracted from its ores by simple metallurgical methods and, moreover, is sometimes found ready for use in its elemental, metallic state. Its ores, because of their abundance and distinctive colours, were easily recognised even by unskilled prospectors. However, copper was not much use for the making of weapons or large structures of any kind until its ability to form a tough alloy with tin was discovered. The discovery of bronze ushered in a great period of human development, the Bronze Age, which lasted for many hundreds of years until the copper alloy was replaced by the more useful iron. In China and India, brass, an alloy of copper and zinc, has played a major role in the lives of millions of people for many hundreds of years. Copper and its alloys are still of importance, not least in the food industry, in the modern world.

CHEMICAL AND PHYSICAL PROPERTIES

Copper has an atomic weight of 63.54 and its atomic number is 29. It has a density of 8.96 and a melting point of 1083° C. It is a tough, though soft and ductile metal, second only to silver as a conductor of heat and electricity. These are the properties which have made copper so valuable to the electrical industry as well as for the manufacture of utensils of many kinds. Copper is resistant to corrosion in the sense that on exposure to air it is superficially oxidised. Copper-sheeted roofs and domes, in the damp, polluted air of cities, take on a patina of green hydroxo-carbonate and hydroxo-sulphate, verdigris, which protects the metal from further corrosion.

Oxidation states of copper are normally 1 and 2. The metal forms two series of compounds, $Cu(I)$, cuprous, and $Cu(II)$, cupric. There is evidence that $Cu(III)$ can also occur in crystalline compounds and some complexes, but, for practical purposes, it is of no significance.

When copper is heated in oxygen it is initially converted to CuO. As the temperature is increased Cu_2O is formed. Copper forms a wide range of cuprous and cupric inorganic salts. The cupric compounds are the most important. Most cuprous compounds are readily oxidised to the cupric form.

The ability of copper to form complexes with amines and other ligands is a distinctive characteristic of the metal and accounts for many of its biological properties. These complexes, which are coloured, also provide the basis for several colorimetric methods for determination of the element, among them the well-known Fehling's test.

PRODUCTION AND USES

Copper ores are widely distributed throughout the world. The most important are malachite, azurite, chalcopyrite, cuprite and bornite. Most are sulphides, but oxides, carbonates, arsenides and chlorides also occur. Many of the ores are mixed with other metals, such as zinc, cadmium and molybdenum.

Copper is relatively easily extracted from its ores by roasting in air to form the oxide and smelting. This is usually followed by refining by electrodeposition.

World production of copper depends very much on industrial demand for the metal. In recent years this has decreased as copper began to be replaced in the electrical and some other industries by cheaper metals and other materials. Even today the primary use of copper, accounting for about half of world production, is in the manufacture of electrical cables and equipment. It is still extensively used for plumbing, though plastic piping has been taking over in recent years. Copper was once extensively used for making cooking and other food processing equipment, but, again, newer metals and other materials have been taking its place. Copper is still occasionally used in sheets for roofing buildings, as well as for covering the hulls of boats. A major use continues to be as a component of alloys, with zinc, tin, cadmium and other metals. The alloy brass, made of copper and zinc, is used in great quantities for domestic food utensils in some parts of the world. Copper-containing bronzes of various kinds, such as phosphor-bronze, are used to make specialist items of machinery and fittings.

Copper salts have important pharmaceutical and agricultural uses. Bordeaux mixture, a fungicide able to prevent blight in potatoes and grapes, was one of the earliest and most important of these to be developed. It is still widely used, as is a variety of copper insecticides. A cupro-arsenate preservative for wooden power poles and fencing stakes is used in many parts of the world as a fungicide.

COPPER IN FOOD AND DAILY DIETARY INTAKE

Copper is present in all foods, both of plant and animal origin. A diet adequate in energy and made up of a variety of different foods, can be expected to provide all the copper that anyone needs for health and complete nutrition. Human copper deficiency is unknown, except in very special cases. Premature infants and patients sustained on total parenteral nutrition or with bowel obstructions, may have an inadequate intake of this essential trace element, but even then it is an unusual occurrence.

Generally the concentration of copper in foods is about 2 mg/kg or less[2]. The best food sources are wholegrain cereals, as well as meat, particularly organ meats, fish and some green vegetables. Certain breakfast cereals, mainly those based on wheat and other grain brans, contain more than 10 mg/kg of copper. Refined cereals and milk products are poorer sources of the metal. Levels in different food groups, as found in a study in the author's laboratory, were (expressed as mean in mg/kg):

> cereals and cereal products: 0.7–3.3; meat/fish: 0.8–2.0; vegetables: <0.2–1.9; milk and milk products: <0.2.

WHO estimates of daily intake requirements are approximately 30 μg/kg body weight for an adult male, with larger amounts of 40 μg/kg for older children and 80 μg/kg for infants[3]. Estimates of daily copper intake by Finnish males of 1.7 mg are somewhat less than the WHO recommendation, at least for larger men, but are adequate for people of smaller body size[2]. The daily intake of 0.75 mg estimated for Australian children in the author's study is also in keeping with the WHO recommendations.

Klevay[4], however, was of the opinion that many modern diets failed to meet nutritional requirements for copper. This he attributed to increasing consumption of refined foodstuffs. He was particularly concerned that excessive refining of foods resulted in an imbalance of intake of zinc and copper which he believed was related to the increased incidence of heart disease in the developed world.

Dietary intake of copper can be influenced, as has been mentioned earlier, to a significant extent by the use of copper utensils in cooking and food processing. It has been shown by the author[5] that while a meal of meat and vegetables, cooked in stainless steel or aluminium, can contribute 0.72 mg of copper to the diet, intake can be more than doubled to 1.83 mg if copper saucepans are used.

Our intake of copper can also be influenced by the quality of the water we consume. Levels of naturally-occurring copper in drinking water may vary considerably, depending on the nature of the soil and rock where it originates. WHO standards for drinking water are 50 μg/litre for copper, but this can easily be exceeded by water which has passed through domestic plumbing where copper pipes and tanks are used. A study of uptake of metals from domestic plumbing systems in an Australian city[5] found that daily consumption of 1.5 litres of water could result in an intake of 80 μg of copper, when water was drawn from the cold tap, but this increased to 12.3 mg when water from the hot system, which used a copper tank and electrical immersion heater, was used.

ABSORPTION AND METABOLISM OF COPPER

The level of copper, like iron, and possibly some of the other essential inorganic nutrients, appears to be carefully regulated by the human body. Absorption takes place mainly in the stomach. The amount of copper absorbed seems to be controlled by the level of copper already in the body in a feedback type mechanism. Normally about 30 per cent of the copper in food is absorbed, but this can increase to about 60 per cent in a low copper diet. On the other hand, when copper ingestion is increased, the percentage absorbed falls.

Several factors, besides total levels of copper in the body and in food, can affect absorption of the metal. Phytate may have an inhibitory effect on absorption. A low molybdenum level in food can result in retention of copper in the body, while excess molybdenum provokes a considerable increase in excretion of copper. Molybdenum-induced copper deficiency in animals is well recognised. Other metals, including iron and zinc, may play a similar antagonistic role to copper[6].

Copper is preferentially stored in organs of the body, including liver, heart, brain and kidney. Some copper is lost in urine, but the main excretion route is via bile in faeces.

Copper is mainly complexed with proteins in blood and tissues. The principal copper protein is ceruloplasmin. This is blue in colour, has a molecular weight of 150 000 and contains 8 Cu(I) and 8 Cu(II) ions. About 3 per cent of the total copper of the body is found in ceruloplasmin. It is manufactured in the liver and is involved in the regulation of levels of copper in the body, and also serves as a transport mechanism for the metal. Ceruloplasmin has an essential role in the oxidation of Fe(II) to Fe(III), an important step in the transport of iron and the manufacture of haemoglobin.

Copper is part of several important enzymes which are concerned with biological oxidation reactions. These include cytochrome oxidase, which is at the terminal stage of the major pathway for cellular oxidation, and tyrosinase, responsible for metabolism of the amino acid tyrosine. This latter enzyme converts the amino acid into melanin, the dark pigment of skin and freckles. Copper is also associated with several amine oxidases, uricase and superoxide dismutase.

While it is clear that a serious copper deficiency could theoretically lead to biochemical defects in cells, and symptoms of copper deficiency are well known in animals, there is no clear evidence that copper deficiency occurs in humans on a significant scale at least. However, copper deficiency has been reported in children fed on cow's milk during recovery from severe diarrhoea[7] and in adults maintained by total parenteral nutrition[8].

A hereditary disease, Menke's 'steely hair syndrome', which is characterised by an inability to absorb adequate copper from the diet, is known to occur in some people. Its symptoms include, as the name suggests, changes in the appearance and structure of the hair, accompanied by serious and widespread defects in cellular metabolism. In contrast, another hereditary disease, Wilson's disease, in which the body accumulates excessive levels of copper in the liver, as well as in the brain, can also occur.

Diseases of humans due either to excessive or deficient accumulation of copper by the body, are paralleled by well-recognised toxicities and deficiencies in animals. Copper-deficient soils occur in many parts of the world and animals and plants living in such areas cannot survive without supplementary supplies of the element. Copper-deficient animals have bone defects, hair damage and anaemia. Protein molecules, in hair as well as blood vessel walls, are poorly cross linked and weak. As a result, hair and wool are deformed and arteries are weakened.

Copper toxicity of animals occurs. A disease of sheep known as 'swayback' is sometimes met with in flocks feeding on copper-contaminated soils. As in humans, copper is an essential trace element in animals. Marginal deficiencies have been observed to occur under certain conditions. Provided it is supplied in an appropriate amount, copper can be used as a growth promoter in some farm animals, especially pigs.

TOXICITY OF COPPER IN THE HUMAN DIET

Copper salts ingested in large amounts will produce toxic reactions in humans as well as in farm animals. The effects are usually reversible, though fatalities have been reported, especially from massive doses. However, copper is an emetic and it is difficult to retain sufficient of a copper salt, such as copper sulphate, to produce fatal effects.

Copper introduced into the body in other ways may have serious consequences. The use of copper piping and connections in a dialysis machine resulted in copper contamination of the dialysate and caused haemolytic anaemia in one case which has been reported[9]. Chronic copper poisoning has also been reported as a result of continued use of water from a hot-water tap[10]. Copper-contaminated milk, due to the use of brass cooking utensils, has been connected with excessive accumulation of the metal in the liver which occurs in Indian childhood cirrhosis[7].

The sensory, as well as the nutritional qualities of food, can be affected by copper. Even a small amount of the metal can act as a catalyst for the oxidation of unsaturated fats and oils, bringing about undesirable changes in taste, odour and colour. These changes, while highly undesirable from the point of view of appearance and consumer acceptability, do not cause a significant increase in toxicity or a decrease in nutritive value of the food.

ANALYSIS OF FOODSTUFFS FOR COPPER

Copper in foodstuffs is readily analysed by flame AAS. The method is accurate, sensitive and shows minimal matrix interference[11]. Digestion of samples is carried out in a nitric–sulphuric–perchloric acid mixture. Digestion is rapid and good recoveries are obtained.

GFAAS is effective for low concentrations of copper, especially in beverages and other fluids. Direct aspiration into the furnace of ready-to-drink beverages is possible. With semi-solid foods or digests containing nanogram levels of copper, extraction into solvents is recommended, with the extract aspirated directly for furnace analysis[12].

IRON

Iron is the second most abundant metal, after aluminium, and the fourth most abundant element in the earth's crust. The earth's core is believed to consist of a mass of iron and nickel and, to judge from the composition of many meteorites that have been analysed, iron is also abundant elsewhere in the environment.

The importance of this metal in the physical world is matched by its importance to human life. Civilisation, as we know it, would not exist without iron. We use more of it than of any other metal. The Iron Age has been followed by the age of steel and technology in which the metal plays an ever increasing role.

The human body, like most other living organisms, has, during the course of evolution, come to depend on iron as a linchpin of existence. The release and utilisation of energy from the food we eat are dependent on the iron-containing enzymes of tissues. Iron is needed for the synthesis of blood pigments, as well as for many other essential activities of cells. Only one group out of all the living organisms that inhabit this world, the anaerobic lactic acid bacteria, appears to possess no iron-containing enzymes.

CHEMICAL AND PHYSICAL PROPERTIES

Iron, which has the chemical symbol Fe after its Latin name, *ferrum*, is element number 26 in the Periodic Table. It has an atomic weight of 55.85 and a density of 7.86. Pure iron is a white, lustrous metal which melts at 1528° C. It is not particularly hard and is ductile and malleable. It has good conducting properties for heat and electricity.

Iron is chemically reactive. In moist air it rapidly oxidises to a hydrous oxide which we call rust. Unlike the patina on copper, rust does not provide permanent protection for the underlying metal since it readily flakes off, exposing fresh metal.

The metal forms two series of compounds, ferrous, Fe(II) and ferric, Fe(III). Other oxidation states also occur, but are not of biological importance. Three oxides are known, FeO, Fe_2O_3, and Fe_3O_4, representing the Fe(II) and the Fe(III) as well as the mixed Fe(II)–Fe(III) oxide which occurs in nature as the mineral magnetite. Iron can be dissolved in acids to form salts. With non-oxidising acids, in the absence of air, Fe(II) salts are obtained. Many iron salts, as well as the hydroxides, are insoluble in water.

PRODUCTION AND USES

Though many soils contain as much as 4 per cent of the metal, iron is not easily extracted from this source. It is obtained from a number of different ores, including haematite, Fe_2O_3, magnetite, Fe_3O_4, limonite, $FeO(OH)$ and siderite, $FeCO_3$. The production of iron is a well- developed and highly-skilled process, involving reduction by smelting with various forms of carbon.

The purposes for which iron is used are so well known and so numerous that they will not be discussed in detail here. Whether in the form of cast iron, or as a component of steel or other alloys, iron plays a major part in the construction of food processing equipment, containers and other utensils used for food. It is through the use of such equipment that most food contamination by iron comes about.

IRON IN FOOD AND BEVERAGES

Iron is present in all foods and beverages, as it is in all biological materials. The levels of the metal in different foods can vary very much, depending on the kind of food and the processes to which it has been exposed.

Concentrations in individual foods and food groups reported in the literature from different countries are similar. The following mean ranges, in mg/kg, have been found in the author's laboratory:

> cereals/cereal products: 6.0–26.5; breakfast cereals (iron forti-
> fied): 20.0–92.0; meat/fish/eggs: 10.3–40.0; milk products:
> <2.0–7.0; vegetables and fruit: <2.0–6.3.

These results are in the same range as those published in such national food composition tables as *McCance and Widdowson*[13] in the UK and the recently published Australian *NUTTAB* computerised food composition data base[14].

Canned fruits, vegetables and other foods may contain elevated levels of iron, compared to fresh produce, due to solution of the metal from the containers. The pH of the foods, as well as the standards used in the canning operations, and several other factors, determine the levels of iron taken up. As has been noted in an earlier section, the presence of nitrate in the canned food is one of the very important determinants of the amount of iron dissolved from the can walls. Some fruits, such as the tropical pawpaw (papaya) can have such high levels of nitrate that they cannot be canned without excessive iron solution from the can. Other fruits, temperate as well as tropical, also present a similar problem for canners. In one study of levels of iron in canned fruit, grapefruit was found to contain 5.3 mg/kg, pears 5.0 mg/kg, and blackcurrants 1300 mg/kg of the metal[15].

The age of a can and the length of time in which it has been in storage are also important factors related to iron levels in canned food. Two studies from Italy present evidence of how important it is for manufacturers and retailers to check the age and storage history of canned products, and the risk that consumers can face if they purchase certain types of old stock of canned goods. The iron content of a variety of canned foods was found to increase by an average of 34.8 per cent after 280 days storage and by 42.8 per cent after 365 days when a maximum concentration of 308 mg/kg of iron was found in the foods. The author of this study[16] recommends that cans should be consumed within the first 1–6 months of storage if excessive intake of iron, and other metals, is to be avoided.

The second Italian study, from the Institute of Pharmaceutical Chemistry, Florence[17], looked at metals in a variety of canned foods, including vegetables, meat, fish and fruit, collected from retail sources. The range of iron concentration in the 98 cans analysed was 2.8–1710 mg/kg. While 74 of these cans had contents less than 50 mg/kg, ten exceeded 200 mg/kg. Two other cans which had been in storage for four years since purchase, which contained anchovies in sauce, contained 1048 and 5800 mg/kg of the metal. A recommendation of the authors was for introduction of legislation in Italy to require date marking of canned foods to help consumers avoid the consequences which could result from consumption of aged cans with high metal concentrations. This is the practice in many countries.

Iron levels in foods can be significantly influenced by the practice of fortification. This is the intentional addition of iron or its salts to certain foodstuffs during their processing. This practice was initially introduced in some countries as a health intervention to prevent iron deficiency resulting from the consumption of refined flour as well as from the generally low level of dietary intake of the metal in the community.

As has been noted by Hallberg in a study of iron deficiency in industrialised countries[18], iron deficiency is one of the few deficiency disorders observed in Western industrialised society. He notes that it may seem paradoxical that a nutritional deficiency can occur in societies where virtually no one is starving. However, we live in a world where energy needs are continually decreasing, because of the replacement by machines of the energy-requiring activities of earlier generations. If excessive body weight is to be avoided, a decrease in intake of dietary energy is essential. The requirements for most nutrients, including iron, are not, however, reduced and adjustments have to be made to the composition of meals to provide the right balance of energy and nutrients. This is done by increasing the nutrient density of foods, that is, the amounts of essential nutrients per unit energy. Unfortunately, in industrialised countries in recent years there is, on the contrary, an increase in the consumption of 'empty calories', in so-called 'junk food' which are rich in high-energy, low-nutrient fat and refined sugar. There is another consideration which is important in relation to the prevalence of iron deficiency anaemia. This is that the amount of dietary iron that is absorbed depends not only on the total intake of iron, but also on the composition of the meals in which the element is consumed. There are two kinds of iron in the diet with respect to the mechanism of absorption, haem and non-haem iron. The haem, or organic iron, in meat is usually better absorbed than non-haem or inorganic iron. The iron in cereals, vegetables and fruits, which usually makes up the bulk of our intake, is non-haem and has a far lower level of bioavailability than iron in meat and other animal products. In addition, other components in the diet can help or hinder iron absorption. Bioavailability can be reduced by phytate, fibre, tannins and oxalate, whereas the presence of ascorbic acid and other organic acids, such as citrate, can enhance absorption of inorganic iron from food[19]. Thus a diet relatively high in nonhaeme iron can, in fact, fail to meet nutritional needs, if the food consumed is also rich in dietary fibre or tannin. It has been shown that one cup of tea can reduce absorption of iron from 11.5 to 2.5 per cent from a meal consisting mainly of bread[20]. On the other hand, the addition of fresh fruit to a meal can increase iron absorption from a mixed meal seven-fold[21].

In the 1940s, when governments were concerned at the effects of rationing and wartime shortages on the community, many health authorities believed that the most appropriate way to combat widescale deficiency of iron, as well as of certain other nutrients, was by the public health measure of compulsory fortification of certain foods. In 1941 the US introduced legislation to require the enrichment of white bread with iron, as well as with the vitamins thiamin, riboflavin and niacin to the level found in whole wheat[22]. The UK adopted a similar programme, as did Sweden and some other countries. It is still a legal requirement in some countries. In others, it is voluntarily practised by certain food manufacturers, especially of breakfast cereals.

Flour was chosen as the vehicle for iron fortification for a number of reasons. A major reason was that it could be argued that during refining wheat flour was deprived of about half of its iron, from about 25 mg/kg in wholemeal (100 per cent extraction rate) to 15 mg in white (72 per cent extraction rate) flour. Fortification, or the addition of iron to white flour, was seen to be simply a restoration of the flour to its original nutritious state. A variety of different forms of iron are used in fortification. Metallic 'reduced' iron is widely employed, as well as ferric ammonium citrate, ferrous sulphate, and ferrous fumerate. The solubility and bioavailability of these forms of iron are not all the same, and they are not all equally effective in meeting the nutritional needs of those who consume them.

In Sweden iron fortification of flour was increased from an initial 30 mg/kg to 65 mg/kg in 1970. Today it is estimated that 42 per cent of total iron intake in the country comes from fortification, and most of this from flour. It is argued by Hallberg that if fortification were discontinued it would result in a reduction of daily iron intake of about 0.35 mg in women who are borderline deficient in the nutrient[23]. An argument was made recently in the US for an increase in iron fortification in foods from the approved level of 28.6 to 36.6 mg/kg. There is currently a growing opinion in Australia, where there is no legal requirement for iron fortification of foods, that it should be introduced to combat widescale iron deficiency anaemia. Some of the pressure for this change is related to the recently established Australian RDI for iron which recommends that pregnant women consume between 22 and 36 mg of iron per day[24]. This level of intake cannot readily be achieved on normal Australian diets and, as the Australian National Health and Medical Research Council acknowledges, requires that pregnant women take a daily iron supplement. It is argued that inclusion of iron-fortified bread in the diet would supply sufficient of the element to meet the high RDI.

There is considerable opposition to legally-required iron fortification of foods. The case against the proposed increase in levels of fortification in the US has been put by Powell and his colleagues[25]. There is good evidence that iron supplementation, or an increased intake through food fortification, may not be desirable for everyone. For instance, an increase in the daily iron intake of infants might saturate the lactoferrin secreted into the intestine and suppress the capability of this protein to contribute to the body's protection against infection[26]. In addition there are two situations in which persons are at risk of excessive storage of iron in the body if they consume an excess of iron[27]. One is in people with refractory anaemia associated with chronic bone marrow failure, such as occurs in thalassaemia. This hereditary defect, which is found most frequently in people of Southern European origin, requires patients to be given regular blood transfusions. The other is in people who have another disorder of iron metabolism known as hereditary haemochromatosis. In both cases, iron overload can occur with accumulation of the metal in organs such as the liver, pancreas and heart. Organ failure, and death, can result. It is of some interest that the consumption of tea has been recommended as a means of reducing excessive iron absorption by patients suffering from thalassaemia. The treatment relies on the iron-binding capacity of tannins in the beverage[28].

A diet high in iron, through contamination of food or beverages, may also result in iron overload, even in the absence of iron-fortified food. This is seen in several parts of the world where iron-rich alcoholic beverages are consumed. In South Africa traditional beers are often made in crude fermentation equipment, resulting in high levels of contamination, by iron and other metals, of the product. Consumption of such beverages in quantity over a long period of time has been shown to cause iron overload and organ failure[29]. This has been also shown to occur in other parts of Africa, including Zambia[30], as well as among some cider drinkers in Normandy. Home-produced, fermented apple beverages have been shown to contain as much as 16 mg/litre of iron. Other alcoholic beverages, including wine, are also known to occasionally have high levels of iron and to contribute to the development of iron overload[31]. It may be surmised that an additional intake of iron, from fortified foods, would have a cumulative effect in such cases of an already existing high dietary intake of the metal.

As has been indicated earlier, the use of iron cooking utensils can contribute significantly to dietary iron intake. A study in the author's laboratory has shown that use of cast iron saucepans to cook a meal consisting of beef, potato and cabbage, could provide an intake of 10.9 mg of iron, compared to 6.1 mg when the food was prepared in aluminium utensils[32]. A study in Papua New Guinea has shown that iron intake can be significantly increased by the use of cast iron pots and that the recent replacement of these utensils by the more readily available aluminium types could result in inadequate dietary iron intake[33]. A similar situation was reported to occur in South Africa where it was postulated that the low incidence of iron deficiency anaemia among black South African women was related to a high iron intake in food cooked in cast iron pots[34].

IRON ABSORPTION AND METABOLISM

As has already been indicated, absorption of iron is a complex process. Metabolism of iron differs from that of other metals in an important respect which is that there is no physiological mechanism for iron excretion. Thus the body maintains its iron balance by controlling absorption of the metal from the gut. Normally only a very small quantity of iron, about 1 mg/day, needs to be absorbed, unless bleeding, as in menstrual flow in women or from other causes, occurs. Absorption is initially into the mucosal cells lining the gut where the metal may be bound into ferritin, an iron-storage protein molecule. This large molecule can accommodate up to 4500 iron atoms, which can be readily mobilised when required by the body. Some of the iron may be transferred directly by a transport protein, transferrin, to other tissues, such as bone marrow where it is used in the synthesis of haemoglobin.

When a state of iron deficiency exists, it appears that this absorption mechanism can be stimulated to increased activity. In contrast, when iron stores are high, the absorption mechanism is slowed down. There is evidence that other metals, such as cobalt and zinc, can compete with iron for this active absorptive pathway. As has been noted above, several other factors besides the level of body stores of iron, affect absorption of the metal. Components of the diet such as phytate, fibre, ascorbic acid, citric acid and animal protein can have either an inhibitory or stimulating effect on iron absorption by forming complexes with the metal. The pH of the gut will also affect absorption. At an acid pH iron is mainly in the more readily absorbed ferrous state.

Iron loss occurs normally from the body through the ageing of cells in the intestinal mucosa. As these are lost from the gut and excreted they carry with them iron which had been incorporated into ferritin. This accounts for the 1 mg/day loss which must be made up by absorption of iron from the diet. Iron released from red blood cells, as these are broken down in the spleen after about 120 days of life, is recirculated in plasma and re-utilised. The kidneys are normally very effective in preventing any loss of this circulating iron.

An adult human contains about 4 g of iron, of which up to nearly 3 g are contained in the two metalloproteins, haemoglobin and myoglobin. Both molecules are able to bind oxygen in a reversible manner. Haemoglobin is packaged within the red blood cells and is used to carry oxygen from the lungs to tissues of the body. Myoglobin is confined to muscle cells where it provides an oxygen reserve.

Other functional haem-proteins also occur in the body. These include cytochrome oxidase and cytochrome P−450 which react with oxygen, and catalase which reacts with hydrogen peroxide. Several others are involved in electron transport mechanisms of cellular respiration. There are also a number of non-haem enzymes, which include an important controller of cell division, ribonucleotide reductase. Between them these other iron enzymes account for about 4 per cent of the functional iron compounds in the body.

Iron deficiency primarily results in anaemia, an inadequate supply of haem-functional proteins. A major consequence is a reduction in the work capacity of the body. Many other associated consequences occur, such as reduced immunity to infection, defects of skin and nails, and mental deterioration.

Iron toxicity can also occur. Consumption of iron tablets by children, who mistake them for sweets, is a common occurrence and one of the most frequent causes of iron poisoning met in many hospitals[35]. Treatment is by oral and parenteral use of iron-chelating compounds, such as desferrioxamine. The agent binds iron still in the gut thus reducing further absorption and also detoxifies iron in plasma and tissues.

Chronic iron toxicity, through consumption of iron-contaminated beverages, resulting in haemosiderosis, has been referred to above. It is unlikely that siderosis would normally result from intake of the relatively high levels of iron sometimes detected in canned food.

ANALYSIS OF FOODSTUFFS FOR IRON

Iron in foods is readily determined by atomic absorption spectrophotometry. Concentrations in most samples generally fall within the working range of flame AAS. Liquid samples can be simply diluted and solid samples dissolved following wet or dry digestion.

For lower concentrations of iron in certain foods, GFAAS is suitable. Digestion in a nitric/sulphuric/perchloric acid mixture has been found to be effective in the author's laboratory. There is no need for preconcentration and recoveries are excellent.

CHROMIUM

Chromium is widely distributed in the earth's crust, though concentrations are generally relatively low, at about 0.04 per cent of the total solid matter. Until very recently little was known about the metal's role in relation to human life. We now know that its low soil concentrations bear little relation to its importance to health. Its toxicity to industrial workers has been recognised for many years[36]. Its essential role in animal life began to be recognised in the 1950s[37]. Progress in establishing the vital role of the element in animal and human metabolism was slow, to some extent because of problems with the determination of the element. As recently as 1977, the late Eric Underwood commented that 'reliable data on the distribution of this element in the animal body are sparse and sporadic', because of the absence of satisfactory methods of analysis for chromium[38]. The situation has changed for the better since Underwood wrote those words and we know a great deal more about the role of chromium in food and human health. However, the problem of accurate analytical procedures has not yet been fully resolved and there is still a dearth of information on concentrations and distribution of the element in some foods and other biological materials.

CHEMICAL AND PHYSICAL PROPERTIES

Chromium has an atomic weight of 52 and is element number 24 in the Periodic Table. It is a heavy metal, with a density of 7.2 and a melting point of approximately 1860° C, the exact temperature depending on the crystal structure of the sample being tested. It is a very hard, white, lustrous and brittle metal and is extremely resistant to corrosion. Chromium can occur in every one of the nine oxidation states, from -2 to $+6$, but only the ground 0, and the $+2$, $+3$ and $+6$ states are of practical significance. The most stable and important oxidation state is Cr(III), which gives a series of chromic compounds, including the oxide Cr_2O_3, chloride $CrCl_3$, and the sulphate $Cr_2(SO_4)_3$. Industrially important Cr(III) salts are the acetate, citrate and chloride.

In higher oxidation states almost all chromium compounds are oxo-forms and are potent oxidising agents. The Cr(VI) compounds include the chromate CrO_4^{2-} and the dichromate $Cr_2O_7^{2-}$. Important compounds of this type include lead, zinc, calcium and barium chromates, as well as sodium and potassium chromates and dichromates.

The relatively unstable Cr(II) chromous ion is rapidly oxidised to the Cr(III) form. Few of the other forms of chromium, such as Cr(IV) or Cr(V), are known or appear to be of biological significance.

PRODUCTION AND USES

Though chromium is present in most soils and rocks, its only important commercial ore is chromite, $FeOCr_2O_3$, which may contain upwards of 55 per cent chromic oxide. This is found in commercial quantities in a few parts of the world and is mined in South Africa and the USSR, with smaller amounts produced in the USA, Zimbabwe and a few other countries. Reduction of the ore with carbon produces ferrochrome, a carbon-containing alloy of chromium and iron. The pure metal is produced from this, either by electrolytic treatment or by conversion into sodium dichromate by treatment with hot alkali and oxygen and reduction by reaction with aluminium.

Chromium has many industrial uses. Of the 400 000 tonnes processed every year in the US, for example, 57 per cent is used in the metallurgical industries, 30 per cent in the production of refractory materials, and the remainder in the chemical industry[39]. Metallurgical uses of chromium are mainly in the production of stainless steel of various kinds, in alloys with iron, nickel and cobalt. Chemical uses of chromium include the production of tanning agents, pigments, catalysts and preservatives. Some chromium is used as an additive in water to prevent corrosion in cooling systems. It is probable that this use of chromium accounts for a significant amount of industrial chromium emission into the atmosphere. The metal is also found in some household detergents, pottery glazes, paper and dyes.

CHROMIUM IN FOOD AND BEVERAGES

Until quite recently, reliable information on levels of chromium in individual foods and intakes in diets was not readily available. This was because of the considerable difficulties which have been experienced in analysing foods for the metal. Chromium is still considered to be one of the greatest challenges to the skill of the analyst[40]. Many of the results published in the past have to be discarded, as they are now seen to have been inflated by contamination during sample preparation, by inadequate background correction during AAS, or through unavailability of CRMs to enable results to be checked against known values. In some cases, even when these difficulties had been overcome, results were suspect because they failed to account for losses through volatilisation during ashing or digestion, or for failure of the sample preparation procedures to release the metal from silica or other complexes.

Accepting that the problem of chromium analysis in food and biological tissues is not yet solved to the satisfaction of all, and that certified reference materials have only recently begun to be developed for the metal, it is possible to quote with some confidence levels of chromium in foods, as reported in recent studies. Smart and Sherlock, of the UK's Ministry of Agriculture, Fisheries and Food, have recently published a comprehensive study of chromium in the UK diet[41]. They give the following figures (mean, with range in parentheses, in mg/kg) for different food groups:

cereals: 0.08 (0.02–0.30); meat: <0.15 (<0.05–0.95); fish: <0.20 (<0.05–0.15); root vegetables: <0.06 (<0.05–0.20); other vegetables: <0.07 (<0.05–0.20); milk: <0.02.

Results from the author's laboratory, where the analytical procedure permitted a lower detection limit than did the UK study, found the following levels of chromium in some corresponding food groups (expressed as mg/kg mean value, ranges in parentheses):

cereals and cereal products: 0.04 (0.001–0.097); meat/fish: 0.027 (0.006–0.051); vegetables: 0.014 (0.003–0.033); whole milk: 0.015 (0.013–0.025).

Differences between the Australian and the UK results may be related to the fact that the latter used foods prepared for the UK Total Diet Study which had been cooked in a variety of different cooking utensils, some of which were made of stainless steel, whereas many of the Australian foods were uncooked. Stainless steel cookware is capable of contributing significantly to the chromium level of food, especially if they are acidic[42].

Some foods are believed to contain exceptionally high levels of chromium. Among these brewers' yeast has been reported to have the highest known concentration of chromium of any foodstuff. Spices such as black pepper came next, with raw sugar also high on the list[43].

Published estimates of daily intakes of chromium have, over the years, generally reflected improvements in analytical techniques, with a downward movement corresponding to the introduction of more sensitive and accurate procedures and instrumentation. For example, the US intake was reported to be 200–290 μg/day by Tipton and colleagues in 1969[44], and 231 μg/day[45] for US schoolchildren by Murthy and colleagues in 1971. A Canadian study[46] reported that daily intake was 282 μg in 1972. In 1980 the chromium intake from a typical Western diet was estimated to be between 50 and 100 μg/day[47]. In the same year a Finnish investigation[2] found that daily intake of chromium was 29 μg. The authors noted that this result was low compared to other estimates and attributed the difference, at least partially, to a possible underestimation of the share of processed foods in the diet. A recent UK study[48] reported that daily intake was 24.5 μg, and a similar US study[43] found an intake of 28 μg. Recent studies in the author's laboratory indicate that the level of chromium in food consumed each day by children is approximately 30 μg, with a range of 16–51 μg. These recent estimates are less than the Estimated Safe and Adequate Daily Intake of 50–200 μg/day proposed by the US National Academy of Sciences. They are also less than the estimates made by Smart and Sherlock, in their UK study, of between 74 (lower bound) and 97 (upper bound) from food, with an additional 1.5–7.5 μg/day coming mainly from water.

The importance of intake from non-food sources of chromium should not be overlooked. Reference has already been made to uptake of the metal from stainless steel cooking utensils. The amount of metal released into food will depend on the type of vessel used, as well as the pH and the temperature of cooking[49]. An example of such pick-up of chromium is given by Jorhem and Slorach[50]. They report that red pickled cabbage, an acidic food, cooked in a stainless steel vat contained 65–72 μg/kg of chromium. The level in fresh vegetables was about 10 μg/kg.

As has been noted in an earlier chapter, the 'passivation treatment' with a chromium compound such as chromic acid or sodium dichromate, which is used to improve lacquer adherence and resistance to oxidation of tinplate, can contribute to the level of chromium in canned foods. It has been calculated that if all the chromium on the inner surface of a 454-g can migrated into the food, it would result in a concentration of about 0.4 mg/kg[51]. Lacquering actually protects the contents of cans from this chromium, as is shown in data from the paper by Jorhem and Slorach[50]. They found that the mean level of chromium in fruit and vegetables in lacquered cans was 0.018 mg/kg, and in unlacquered cans 0.090 mg/kg, while the fresh food had only 0.009 mg/kg.

Though water may contain some chromium, especially if it is affected by emission from industry, it is unlikely to make a significant contribution to chromium intake. The presence of Cr(III) in drinking water is unlikely because of the low solubility of the hydrated Cr(III) oxide. The more soluble Cr(VI) may occur in water especially in the vicinity of industries which cause pollution. The WHO standard for drinking water[52], which is also adopted by the EEC[53], is a maximum of 50 μg/litre of Cr(VI). Chromium concentrations in drinking water in some European cities are believed to range from 1.0 to 5.0 μg/litre, with a mean of 2.0 μg/litre. As Smart and Sherlock[41] have indicated, this would mean, on the basis of a daily consumption of 1.5 litres of water, a daily intake of between 1.5 and 7.5 μg.

ABSORPTION AND METABOLISM OF CHROMIUM

The mechanism of absorption of chromium from food in the intestine is not well understood. The amount absorbed seems to depend on the chemical form present. Only 0.5–1.0 per cent of trivalent chromium and possibly about 2 per cent of chromates will enter the bloodstream. However, it has been reported that between 10 and 25 per cent of the chromium in brewer's yeast is absorbed by rats[54]. This is believed to reflect the fact that chromium in yeast occurs as an organic complex, the 'glucose tolerance factor' (GTF) which will be discussed later. Chromium absorption is inhibited by the presence of certain other metals in food. Iron, manganese, vanadium, and possibly zinc, may compete with chromium for a common pathway from the intestine and thus have mutually inhibitory effects on uptake[55].

Chromium appears to accumulate principally in the kidneys, spleen, testes and bone and other organs. Absorbed chromium is mainly excreted in urine, with a very small amount in bile through the faeces. The chromium content of all tissues is reported to decrease with age, except in the case of the lungs in areas where environmental contamination is high.

Chromium is believed to be an essential trace nutrient in man and other animals[56]. Its essentiality has only been recognised since the late 1960s and we still know relatively little about its biological functions. Of particular importance appears to be its role in relation to carbohydrate metabolism, in which it is closely related to insulin. Chromium is believed to be an integral part of a dinicotinic acid–glutathione complex, the so-called glucose tolerance factor. GTF potentiates the action of insulin and has been shown to be present in significant amounts in brewers' yeast. Chromium may also have a role in lipid metabolism. There is some evidence that chromium deficiency may result in impaired glucose tolerance and may contribute to heart disease. However, the evidence for this is not yet well established[48].

There is evidence that Cr(III) is capable of binding to nucleic acids and thus could cause abnormal synthesis of genetic materials in cells[57]. The deleterious effects on health of industrial exposure to chromium and its vapour and salts are well known. As has been noted by Burrows[39], chromium is important in medicine because of its widespread use, toxicity, carcinogenic effect, and its high sensitisation index in low concentrations. There is no evidence that the chromium normally present in foods is a danger to health, whether the metal is present naturally in food or is a result of the use of stainless steel utensils. Indeed, adventitious chromium from such non-food sources may have health advantages, as has been mentioned already.

ANALYSIS OF FOOODSTUFFS FOR CHROMIUM

In spite of recent developments in trace metal analytical techniques, chromium remains an extremely difficult element on which to obtain reliable data. The low levels of the element normally found in foods and other biological samples, and the acute danger of contamination, contribute greatly to the analytical difficulties[40]. These problems have been commented on by Veillon[40] and have been studied extensively in the author's laboratory[58].

Sample preparation presents particular difficulties. The use of stainless steel equipment for homogenising or sample taking must be avoided. All preparation should be carried out in a clean room, preferably using a laminar flow or other suitable work station[59].

Dry ashing has been reported to be suitable for sample preparation, provided temperature is kept at 460° C[40]. Others have reported losses of chromium through volatilisation when this technique is used[60]. Acid digestion has been found to be more satisfactory than dry ashing in the author's laboratory. Use of a nitric–sulphuric acid digestion mixture was found to give good reproducibility of results, except when food samples contained high levels of carbohydrates and fat. In these cases the problem was overcome by drop-wise addition of nitric acid.

Problems were also experienced in recovery of chromium after acid digestion of foods which contained silicates. Similar difficulties have been reported by other investigators[61]. The addition of hydrofluoric acid was found to overcome this difficulty which was experienced only in foods of plant origin, including brewers' yeast.

Atomic absorption spectrophotometry, with electrothermal atomisation and effective background correction, is widely used for estimation of chromium in food digests. As has been pointed out by Versieck and Cornelis[62], one of the main problems in applying GFAAS to chromium determination is background correction. The deuterium lamp is of little use because its intensity at the 357.9 nm line is very low compared to the hollow cathode lamp and will not eliminate the non-specific absorption. A high-intensity tungsten-iodide lamp has been used to overcome this problem by some investigators[63]. The author has used Zeeman mode background correction with good effect and achieved reduced background absorption and improved sensitivity.

Where available, NAA, since it avoids extensive sample preparation, and therefore much of the possibility of contamination, has been shown to be a suitable and accurate method for the determination of chromium in food.

MANGANESE

Manganese is element number 25 in the Periodic Table, with an atomic weight of 54.94. Its chemical and physical properties are, in many respects, similar to those of iron, which it immediately precedes in the first transition series. However, it is harder, more brittle and less refractory than iron, with a melting point of 1247° C. Like iron, it is a reactive metal, which dissolves readily in dilute, non-oxidising acids. It ignites in chlorine to form $MnCl_2$. It reacts with oxygen at high temperatures, producing Mn_3O_4. Manganese also combines directly with boron, carbon, sulphur, silicon and phosphorus.

Manganese has several oxidation states, the most important of which, from a practical point of view, is the divalent, Mn(II). This forms a series of manganous salts with all the common anions. Most of these are soluble and crystallise as hydrates. Mn(II) also forms a series of chelates with EDTA, and other agents.

The chemistry of Mn(III) is not extensive. In aqueous solution it is unstable and is readily reduced to Mn(II). Similarly Mn(IV) compounds are of little practical importance, except for the oxide MnO_2, pyrolusite, which occurs in nature and is one of the chief ores of the metal.

Mn(VI) is found only in the manganate ion $(MnO_2)^{2-}$. Similarly Mn(VII) is best known in the permanganate ion $(MnO_4)^-$. Potassium permanganate is a common and widely used compound. It is a powerful oxidising agent and has extensive pharmaceutical as well as chemical applications. It is traditionally known to many, as a wine-red solution, under the name of Condy's fluid.

PRODUCTION AND USES

Manganese is relatively abundant, making up about 0.1 per cent of the lithosphere. It is the twelfth most common element. Of the heavy metals, only iron is more abundant than manganese. Ores of manganese occur in a number of substantial deposits, especially as pyrolusite, the Mn(IV) oxide. Major commercially-exploited deposits occur in Africa, the USSR, Canada and Australia. Production of the metal is by roasting of the ore followed by reduction with aluminium. Manganese has several important industrial applications. It is used in steel and other alloys, in the manufacture of electrical accumulators and in a variety of applications in the chemical industry. In recent years it has been introduced in some countries as a replacement for organic lead additives in petrol. Manganese is also used in pigments, glazes, domestic and pharmaceutical preparations and other products.

MANGANESE IN FOOD AND BEVERAGES

Manganese is present in all animal and plant tissues, in water and in dust, as well as in all human tissues[64]. Until recently little reliable information was available on the manganese content of different foods or on dietary intakes of different population groups[65], but the situation has been changing as the significance of manganese for health is better recognised and improved analytical techniques become available.

The level of manganese in comparable foods reported by different workers in several countries appears to be relatively similar, but concentrations in different types of foods show wide variations. A UK report[64] provides the following data on levels in different food groups (expressed as mg/kg, mean value, with the range in parentheses):

cereals: 6.77 (2.4–14.0); meat/fish: 0.72 (0.45–1.1); milk: <0.10; vegetables: 1.43 (0.5–2.2).

Similar values have been found in the author's laboratory. Milk had 0.05 mg/kg (range 0.04–0.07 mg/kg). As in the UK study, cereals and especially wholemeal products, were found to be highest in manganese, with the lowest concentrations in meat.

An interesting, and in the national context, important finding in the UK study was that tea can be a major source of manganese in the diet. Dry tea was found to contain between 350 and 900 mg/kg, and tea, as consumed, 7.1–38.0 mg/kg of manganese.

It was estimated that daily intake of manganese in the UK diet was 4.6 mg. These results are close to estimates of daily intakes published for many other countries. It is possible that a diet in which wholemeal cereals and tea are not consumed could result in a low intake of manganese. Some concern has been expressed that this may be the case in modern society and that the daily intake of manganese may be less than adequate for some people. Processing causes a considerable loss of manganese in cereals and in other foods[66]. Wholemeal flour contains about 25 mg/kg of manganese while white flour has about 7. Manganese is reduced from about 0.4 to 0.05 mg/kg of the metal when it is refined. Polished rice contains 1.53 mg/kg compared to 2.8 mg/kg in the whole grain.

ABSORPTION AND METABOLISM OF MANGANESE

Not a great deal is known about the absorption and metabolism of manganese. Uptake from foods is low, with some considerable variation of absorption levels between individuals. In a recent study[65] Davidson and colleagues found an average retention rate of approximately 3 per cent, with a range of 0.6–9.2 per cent in men. Absorbed manganese leaves the bloodstream quickly and is concentrated largely in the liver and other organs. Manganese can cross the blood–brain barrier and the placenta. Excretion of manganese is mainly in bile with faeces. Only a small amount appears in the urine.

Manganese is an essential nutrient for humans and other animals[67]. It is a component of several important cellular enzymes such as pyruvate decarboxylase, arginase and possibly others involved in nucleic acid metabolism. Other enzymes, such as succinic dehydrogenase, require manganese as an activator. Manganese is also a component of one of the superoxide dismutases which are important in protecting cells against oxidative damage by free radicles. The element appears to be essential for normal bone and cartilage formation. Some well-defined deficiency diseases of manganese, which result in skeletal deformities, are known in animals.

Manganese has long been recognised as a toxic element to industrial workers and miners. A 'metal fume fever' in miners and others who use the metal, affects the central nervous system and can lead to acute psychic and neurological symptoms. This has been given the name 'manganese madness'. The symptoms of manganese toxicity are very similar to those of Parkinson's disease, resulting from degeneration of the cerebellum and disruption of nerve function in the area. Treatment with dopamine alleviates symptoms of both diseases.

Until very recently it was believed that while fumes and dust of the metal, in industry, might lead to manganese toxicity, there was no danger from ingestion of the metal in foods. Wenlock and colleagues in the UK study cited above[64] noted that 'Mn is one of the least toxic of all elements and intakes of as high as 500–2000 mg/kg food appear necessary even to retard growth in farm and experimental animals'.

The belief that manganese in food is not potentially toxic to humans has been challenged by reports published by Florence[68] and others[69] since the mid 1980s on the neurological disorders suffered by the inhabitants of Groote Eylandt, an island in the Gulf of Carpentaria in Northern Australia. More than 1000 aboriginal people of the Angurugu tribe live there on top of one of the world's largest high grade manganese ore deposits.

The people live close to open-cut mines and crushed ore piles. A layer of black pyrolusite dust covers everything. Soil on which vegetables are grown contains upwards of 103 000 mg/kg of manganese. Locally-grown bananas contain 31 mg/kg, yam 720 mg/kg and another local food, paspalum, 240 mg/kg of the metal. It has been estimated that a local diet could contribute between 100 and 200 mg/kg each day of manganese[70]. The ESADI for manganese in the US is 2.5–5.0 mg/day for adults[47]. Up to 2 per cent of the local aboriginal population have been found to be affected by motor-neurone defects and cerebellar dysfunctions. A distinctive symptom of the condition is an awkward, high-stepping gait, which has earned those afflicted the local name of 'bird people'.

The precise cause of the illness has not yet been determined. Some evidence has been presented at a conference of experts held to discuss the problem[71] of a possible genetic predisposition to the illness in some susceptible individuals which is triggered by the high environmental level of manganese. The likelihood of chronic manganese poisoning, through a high intake in food and water, as well as by inhalation of dust and absorption from the lungs, is considerable.

ANALYSIS OF FOODSTUFFS FOR MANGANESE

Manganese is relatively easily determined in food by acid digestion, followed by GFAAS. For low levels of concentration, background correction is essential. Normal precautions to prevent contamination of samples are required.

In the author's laboratory a nitric–sulphuric–perchloric acid mixture has been found to be effective for preparing foods for manganese determination. Flame AAS was sufficiently sensitive for foods with relatively high levels of the metal. Zeeman mode background correction was used with excellent results for lower levels of concentration.

COBALT

Cobalt is one of the most fascinating of the essential trace metals, with several unusual facets. It is a relatively rare element, making up only about 0.001 per cent of the lithosphere. It occurs in animals, including man, in minute amounts, and the daily intake of the human body is not much more than 0.1 μg. It is an integral part of cobalamin, or vitamin B_{12}, of which the body normally contains not much more than 5 mg. However, if there is less than this, the body can develop pernicious anaemia, a fatal illness. Its role in vitamin B_{12} appears to be the sole function of cobalt in the human organism. Intake of the element to meet our requirements must be as the preformed vitamin. The metal itself is not enough. Indeed, in more than minute amounts it is toxic.

CHEMICAL AND PHYSICAL PROPERTIES

Cobalt has an atomic weight of 58.9, and an atomic number of 27. It has a density of 8.9 and is a hard, brittle, bluish-white metal, with a melting point of 1490° C. The metal is fairly unreactive and dissolves only slowly in dilute acids. Like the other transition metals, cobalt has several oxidation states, but only states 2 and 3 are of practical significance. Co(II) forms a series of simple and hydrated cobaltous salts with all common anions. Few cobaltic, Co(III), salts are known, since oxidation state 3 is relatively unstable. However, numerous stable complexes of both the cobaltous and the cobaltic forms of the element exist.

PRODUCTION AND USES

Cobalt is found in association with other elements, especially arsenic. Smaltite, $CoAs_2$, and cobaltite, CoAsS, are two important ores. However, the chief commercial source of the element is the residue left after the refining of nickel, copper and lead from which cobalt is extracted by electrolysis.

The principal use of cobalt is for the production of high-strength alloys.

The metal is also used to make permanent magnets, in the manufacture of glass and glazing for pottery, as well as in the pharmaceutical industry.

COBALT IN FOOD AND BEVERAGES

Levels of cobalt are normally very low in foods. Average daily intakes in the US were estimated to be in the region of 150–600 μg in 1966[72]. However, more recent estimates have reduced this figure considerably. Varo and Koivistoinen[2] have calculated that daily intake of a Finnish adult is a total of 13 μg. Of this, 4.3 μg comes from cereals, 3.6 μg from vegetables, less than 1.0 μg from meat and fish and 1.5 μg from dairy products. The Finnish study also provided information on levels of cobalt in individual food groups, as follows (expressed as average concentrations in ng/g, dry weight):

> cereals: 25; meat: 7.8; fish 32; dairy products: 12; vegetables: 46; other foods: 21.

Higher than normal intakes of cobalt are reported to have occurred in beer drinkers because of the use of cobalt salts as an additive to improve the foaming qualities of the beverage[73]. This practice is no longer in use following fears that excessive consumption of cobalt in this way, up to the equivalent of 8 mg of cobalt sulphate each day, caused an epidemic of cardiac failure in heavy drinkers. An investigation produced evidence that the cobalt alone was not responsible for the heart failure, but that two other factors, a high alcohol intake and a dietary deficiency, either of protein or of vitamin B_1, played a major part in the incident[74].

ABSORPTION AND METABOLISM OF COBALT

Reports by different investigators on levels of absorption of cobalt present a confusing picture, with estimates ranging from 5 to 45 per cent[75]. The variation may, in part, be due to nutritional and other factors which affect levels of absorption. These may, for example, be increased when a state of iron deficiency exists in the body. The chemical form of the element can also affect uptake from food.

The whole body content of cobalt in an adult man is little more than 1 mg[76], with perhaps a fifth of this stored in the liver. Excretion of cobalt is mainly in urine. As has already been noted, cobalt is required by man only as a component of vitamin B_{12} and this vitamin must be taken into the body in a preformed state. Only micro-organisms, and not humans, can incorporate the inorganic metal into the vitamin, so that it is the vitamin and not the metal that is required in the diet.

No functions for cobalt in human nutrition, other than as an integral part of vitamin B_{12}, have been clearly established. The vitamin plays a very important role in the body, especially in cells where active division is taking place, such as in blood-forming tissues of bone marrow. If there is a deficiency of the vitamin, abnormal cells known as megablasts develop. This leads to potentially fatal macrocytic anaemia. In addition to the anaemia, the nervous system is also seriously affected.

There is some evidence that cobalt is necessary for synthesis of the thyroid hormone. A Russian report noted an inverse relationship between cobalt levels in food, water and soils in certain areas and the level of incidence of goitre in animals and humans[77].

ANALYSIS OF FOODSTUFFS FOR COBALT

Because of low levels of concentration in most foods, the determination of cobalt requires considerable care and analytical skills. Atomic absorption is generally the method of choice. Flame AAS is relatively selective, but insufficiently sensitive. Graphite furnace AAS with background correction is suitable in most cases. Wet digestion in an acid mixture followed, if necessary, by a preconcentration solvent extraction, can be employed. Sensitivity can be improved considerably by the use of pyrolytically-coated graphite tubes[78].

NICKEL

CHEMICAL AND PHYSICAL PROPERTIES

Nickel is element number 28 in the Periodic Table with an atomic weight of 58.71. It is a tough, silver-white metal with a density of 8.9 and a melting point of 1453° C. Nickel is very adaptable to the needs of fabricators. It can be drawn, rolled, forged and polished, and has good heat and electrical conductivities. Though highly resistant to attack by air and water, it can be readily dissolved in dilute acids. Like the other transition metals, nickel has several oxidation states, including Ni(0), Ni(I) and Ni(II), but the last is the only state that is common. The chemistry of Ni(II) is complex, because of its tendency to form complexes. It forms a number of binary compounds with various non-metals such as carbon and phosphorus. Divalent nickel forms an extensive series of simple compounds also, many of which are green in colour.

PRODUCTION AND USES

Nickel occurs naturally in combinations with arsenic, antimony, sulphur and some other elements. Its principal ores include pentlandite, kupfer–nickel and garnierite, a magnesium-nickel silicate, as well as certain varieties of the iron mineral pyrrhotite, which can contain as much as 5 per cent nickel.

Elemental nickel has been found, as an alloy with iron, in some meteorites. There is evidence that the earth's core may contain considerable quantities of the metal.

Extraction and refining of nickel from its ores is complicated. Usually the ore is first converted to Ni_2S_3 and then to the oxide NiO by heating. The oxide is reduced to the metal by hydrogen. Pure, high-grade nickel is produced by the Mond process in which the crude metal is reacted with carbon monoxide to form volatile $Ni(CO)_4$. Further heating, to 200° C, produced 99.99 per cent nickel.

Nickel's major use is in the production of high quality, resistant alloys with iron, copper, aluminium, chromium, zinc and molybdenum. Nickel-containing steels are strongly corrosion-resistant. These steels are used in the manufacture of some food processing equipment and account for some of the metal in the diet. There are said to be at least 3000 different alloys of nickel[79]. These are used extensively in storage batteries, automobile and aircraft parts, electrodes, coinage and many other ways.

Nickel is also used in the chemical industry, for example as a catalyst in the hydrogenation of edible oils and fats. Salts of the metal are also used in the manufacture of certain pigments in lacquers, paints, pottery and cosmetics. Nickel sulphate is used as a mordant in dyeing and in printing of fabrics.

NICKEL IN FOOD AND BEVERAGES

Nickel occurs in small amounts in most soils, and plants may contain between 0.5 and 3.5 mg/kg of the metal[80]. It can be detected in most tissues of the human body in small amounts[81]. Until relatively recently little accurate information was available on levels of nickel in foods and diets. This deficiency is now being remedied as improved analytical techniques are developed for the metal. A recent UK study[79] has provided information on which the following summary of levels of the metal in different food groups is based (data expressed as means, in mg/kg, with ranges in parentheses):

Cereals/cereal products: 0.16 (0.10–0.30); meat/fish: <0.12 (<0.05–1.1); vegetables: 0.11 (<0.05–0.30); milk: < 0.02.

On the basis of such findings, the mean daily intake of nickel in the UK diet was estimated in 1984 to be between 0.17 mg (upper bound) and 0.15 mg (lower bound). These results are close, or equal, to estimates of average intake of the metal in Finland (0.13 mg/day)[2], and in the US (0.17 mg/day)[82]. However, as has been noted in another study, such daily intakes can be increased greatly if the diet contains foodstuffs which have higher than average concentrations of nickel[83].

Certain foodstuffs naturally contain high levels of nickel. One of these is tea. Tea leaves were found in the UK study[79] to contain between 3.9 and 8.2 mg/kg of the metal, and, instant tea, between 14 and 17 mg/kg. This study also found that certain herbs contained up to 6.0 mg/kg of the metal. It is unlikely that either commodity will make a major contribution to total dietary intake of nickel.

A Dutch study[84] found that while most foods contained less than 0.5 mg/kg of nickel, two foods, nuts and cacao products, contained considerably more. Samples of cacao, used in the manufacture of cocoa and chocolate, had up to 5 and 10 mg/kg. A US study[85] found that most foods contained less than 0.4 mg/kg of nickel. Higher levels (above 1 mg/kg) were found in nuts, legumes, products containing chocolate, grain products and canned foods.

Contamination of food by nickel during processing and storage in metal containers has been studied by a number of investigators. The contribution to dietary intake due to migration from metal cookware was estimated to be 0.1 mg/day[79]. A small contribution can be made also to nickel intake by uptake from cans. However, the significance of this is doubtful. There is some evidence of pick-up of the metal by milk from stainless steel containers[86] and by wine from similar vessels[87], but again the significance for total dietary intake is minimal. Nickel may also be accumulated in food crops grown on soils contaminated by industrial effluents of sewage sludge, but the amount of excess uptake is not great[88].

ABSORPTION AND METABOLISM OF NICKEL

Nickel is poorly absorbed from food and beverages. Between 3 and 6 per cent of dietary intake is believed to be retained in the body. Excretion is mainly in faeces, with a small amount in urine. There does not appear to be accumulation in any particular organ, but the metal is fairly evenly distributed among the tissues of the body[89]. Nickel is recognised as an essential nutrient for animals[90]. It is also believed to be essential for man[3]. The enzyme urease is a nickel-containing enzyme. Several other enzymes are activated, though not exclusively, by the metal. These include carboxylase, trypsin and acetyl coenzyme A.

Dietary nickel is, apparently, non-toxic to man. However, cancer of the respiratory tract and dermatitis occur in workers in nickel refineries[91]. There has been a growing interest in possible toxic effects of nickel in foods, especially in relation to allergenic effects in recent years[92].

ANALYSIS OF FOODSTUFFS FOR NICKEL

The method of choice for nickel analysis is graphite furnace atomic absorption spectrophotometry, with adequate background correction. The Zeeman mode is most suitable for this purpose. As Versieck and Cornelis note, the Zeeman method is superior to the more classical deuterium background correction system and allows for determination of the metal after minimal sample preparation and without tedious enrichment extractions[62].

Sample preparation may be by digestion in a nitric–sulphuric–perchloric acids mixture. If necessary, nickel may be extracted from the digest for preconcentration into MIBK with APDC at pH 2–4.

MOLYBDENUM

Though molybdenum has been known to science for two centuries, following its isolation by Hjelm in 1782, its importance in animal and human life has not been appreciated until very recently[93]. In 1930 Bortels found that it acted as a catalyst for nitrogen fixation in a bacterium. In 1943 its role as a constituent part of the enzyme xanthine oxidase in rat liver was recognised. In 1967 its essentiality for humans was established when the pathology of Mo-deficiency was recognised in human patients[94]. Molybdenum has not yet attracted much attention from health authorities or legislators, either as a nutrient or a contaminant of diets. However, an ESADI has been established for the metal in the US and it is likely that it may be added to RDIs in several countries as more becomes known about its role in health.

CHEMICAL AND PHYSICAL PROPERTIES

Molybdenum has an atomic weight of 95.94 and is number 42 in the Periodic Table of the elements. It is dull, silver-white in colour and is a malleable, ductile metal, with a high melting point of 2610° C. Molybdenum metal is strongly resistant to acid attack. It undergoes oxidation when heated in air or oxygen, but at lower temperatures is not easily oxidised.

The chemistry of molybdenum is complex. Like other transition metals, molybdenum has a number of oxidation states, from -2 to $+6$. However, only the higher oxidation states are significant in practice. The lower states occur mainly in organo-metal complexes.

The most stable oxidation state is Mo(VI). The trioxide, MoO_3, is a stable, white solid. It can be dissolved in alkaline solutions to form molybdates $-MoO_4$. These tend to polymerise into polymolybdates of the form $X_6Mo(Mo_6O_{24}).4H_2O$. Numerous complexes of Mo(V) are known, including cyanides, thiocyanates, oxyhalides and a variety of organic chelates.

Molybdenum can form alloys with other metals. The chemistry of these, like that of its organic compounds, is extensive and complex.

PRODUCTION AND USES

Molybdenum is one of the rarer elements in the lithosphere. It occurs at low levels in soils normally, though some areas of high soil Mo are known, especially over certain types of shale deposits. A number of ores of the metal are known, the most important of which is molybdenite, MoS_2. This occurs in several parts of the world but the only major deposit is in Colorado in the US. The metal is extracted from the ore, which is initially crushed and subjected to a flotation process, by roasting to form MoO_3. The trioxide is then reduced to the metal by treatment with hydrogen.

The chief use of molybdenum is in the production of steel alloys. These are extremely hard and strong and are used extensively for the manufacture of machinery and military hardware. Molybdenum is also used widely in the chemical industry, and in other manufacturing, as a catalyst, in paints and pigments and in the production of pharmaceutical and agricultural products.

MOLYBDENUM IN FOOD AND BEVERAGES

A number of recent studies have been published to fill the gap which, until quite recently, existed in our knowledge of levels of molybdenum in diets. Molybdenum occurs in all foods, but at very low levels, usually less than 1 mg/kg, as the following data indicate. These are based on results from the US Total Diet Study and represent levels in foods as purchased and consumed by the normal community in several US cities[85]. (They are expressed as means, in μg/kg, with ranges in parentheses):

cereals/cereal products: 300 (26–1170); meat/fish: 89 (4–1290); milk/dairy products: 46 (19–99); vegetables: 51 (5–332).

The major sources of molybdenum were found in this study to be cereals (31–39 per cent of total intake). Milk and dairy products made a major contribution to intake by infants (28 per cent of total). Though in general vegetables did not contain high levels of molybdenum, legumes did make a significant contribution to overall intake of the metal, with up to 19 per cent of intake provided by this one type of vegetable.

Results from an earlier US study[95] of molybdenum in the diet are close to those reported in the Total Diet Study. Based on these figures, it was estimated that the average daily intake of an adult in the US is between 120 and 240 μg. This is a little higher than the 50–126 μg/day estimated in the later US study, but lower than several earlier estimates from several countries which range from about 250 to 1000 μg/day[85]. As has been noted in earlier sections of this book, more recent estimates of intakes of several metals which occur at very low levels in foods, are often considerably lower than earlier published data would suggest. This reflects improvements in analytical instrumentation and procedures.

There does not appear to be any particular foods or types of foods which, in the absence of extrinsic factors, naturally have high levels of molybdenum. However, environmental pollution, from natural or man-made sources, can lead to high levels of the metal in plants. It has been reported that residents of molybdenum-rich regions of Armenia and India, whose food is mainly obtained locally, may have abnormally high intakes of up to 15 mg/day[93]. The use of sewage sludge or molybdenum-containing fertilisers, can result in increased soil levels of the metal, with consequent increased uptake by plants and animals[96]. In the case of cattle, a distinctive toxic condition known as 'teart disease' occurs where pastures have molybdenum levels up to 100 mg/kg, compared to the normal range of 3–5 mg/kg (dry weight). In addition, such pastures can result in secondary copper deficiency in cattle as a result of competition between the two metals for absorption[97].

ABSORPTION AND METABOLISM OF MOLYBDENUM

Molybdenum in food is readily absorbed by the body, possibly by a passive mechanism. More than 80 per cent is absorbed in the stomach and the remainder in the small intestine[93]. Retention rates are low and most of the absorbed metal appears to be excreted in urine within a few hours of uptake.

Molybdenum is a constituent of several important cellular enzymes, including xanthine oxidase, necessary for the formation of uric acid, aldehyde oxidase and sulphite oxidase. The latter is essential for the detoxification of sulphite arising from the metabolism of sulphur-containing amino acids, from ingestion of bisulphite preservatives and from inhalation of sulphur dioxide. It is believed that in all Mo-containing enzymes, the metal is part of a complex called the 'molybdenum cofactor'. The chemical structure of this is not known, but it is believed to be a reduced pterin grouping. A consequence of this need in the body for incorporation of Mo into a complex structure is a potential for genetic abnormalities. There is increasing evidence that inborn errors of metabolism involving sulphite oxidase and xanthine oxidase activities occur, with serious health consequences for the victims[93].

Metabolism of the metal is closely related to that of sulphur, and, as we have seen above, of copper. Sulphur and sulphur compounds are known to be capable of limiting molybdenum absorption and retention in farm animals. Sulphur is, in fact, used to treat 'teart disease', molybdenum toxicity, in cattle.

The essentiality of molybdenum for animals, including humans, has been argued mainly on the basis of its presence in all tissues and its role as a prosthetic group in the enzymes mentioned above. These enzymes are important for the catalysis of oxidation–reduction reactions in cells. In spite of such strong evidence, it was difficult to demonstrate, unequivocally, a dietary requirement for molybdenum in animals, and, still more so, in man. However, the recent report of a case of molybdenum deficiency in a patient on long-term total parenteral nutrition, has put the matter beyond doubt[98]. In recognition of its essential role, molybdenum has been given an ESADI by the National Research Council in the US[47]. This is 150–500 μg/day for adults, 50–100 μg/day for children and 40–80 μg/day for infants.

While molybdenum toxicity is well recognised in animals, it has not been clearly established in man. There is a report of a possible connection between the consumption of molybdenum-rich locally grown foods and a high incidence of gout in Armenia[99]. It is suggested that chronic exposure to molybdenum causes an increase in xanthine oxidase activity, leading to increased uric acid production, and thus gout in local residents.

ANALYSIS OF FOODSTUFFS FOR MOLYBDENUM

The determination of molybdenum in foods presents some considerable problems because of its generally very low levels of concentration. However, the application of standard analytical procedures and precautions against loss or contamination can lead to satisfactory results. Use of graphite furnace atomic absorption spectrophotometry, with deuterium lamp or Zeeman mode background correction, is generally the method of choice for such analysis. Acid digestion is recommended and, if necessary, subsequent extraction and concentration of the element.

REFERENCES

1. COTTON, F.A. and WILKINSON, G. (1966). *Advanced Inorganic Chemistry. A Comprehensive Text* (Interscience, New York).
2. VARO, P. and KOIVISTOINEN, P. (1980). Mineral composition of Finnish foods XII. General discussion and nutritional evaluation. *Acta Agric. Scand., Suppl.* 22, 165–70.
3. WORLD HEALTH ORGANISATION (1973). *Trace Elements in Human Nutrition. Tech. Rep. Ser. No.* 532 (WHO, Geneva).
4. KLEVAY, L.M. (1975). Ratio of zinc to copper in diets in the United States. *Nutr. Rep. Intn.*, 11, 237–42.
5. REILLY, C. (1985). The dietary significance of adventitious iron, zinc, copper and lead in domestically prepared food. *Food Add. Contam.*, 2, 209–15.

6. BOYNE, R. and ARTHUR, J.R. (1986). Effects of molybdenum or iron induced copper deficiency on the viability and function of neutrophils from cattle. *Res. Vet. Sc.*, 41, 417–19.

7. LAHEY, M.E. and SCHUBERT, W.L. (1975). Copper availability from milk. *Am. J. Dis. Child.*, 93, 31–4. TANNER, M.S., BHAVE, S.A., KANTARJIAN, A.H. *et al.* (1983). Early introduction of copper-contaminated animal milk feeds as a possible cause of Indian childhood cirrhosis. *Lancet*, ii, 992–5.

8. KARPEL, J.T. and PEDEN, W.H. (1972). Copper deficiency in long term parenteral nutrition. *J. Pediat.*, 80, 32–6.

9. IVANOVICH, P., MANZLER, A. and DRAKE, R. (1969). Copper toxicity from copper contaminated water in renal dialysis. *Trans. Amer. Soc. Art. Intern. Organs*, 15, 316–20.

10. SALMON, M.A. and WRIGHT, T. (1971). Chronic copper poisoning presenting as pink disease. *Arch. Dis. Childhood*, 46, 108–10.

11. EVANS, W.H., DELLAR, D., LUCAS, B.E. *et al.* (1980). Observations on the determination of total copper, iron, manganese, and zinc in foodstuffs by flame atomic-absorption spectrophotometry. *Analyst*, 105, 529–41.

12. MATOUSEK, J.P. and STEVENS, B.J. (1971). Biological applications of the carbon rod atomiser in atomic absorption spectrophotometry. *Clin. Chem.*, 17, 363–8.

13. PAUL, A.A. and SOUTHGATE, D.A.T. (1978). *McCance and Widdowson's The Composition of Foods.* 4th edn, revised (HMSO, London and Elsevier/ North Holland, Amsterdam).

14. DEPARTMENT OF COMMUNITY SERVICES AND HEALTH (1989). *NUTTAB89: COMPOSITION OF FOODS, AUSTRALIA COMPUTER PROGRAM* (Australian Government Publishing Service, Canberra, ACT).

15. CROSBY, N.T. (1977). Determination of heavy metals in food. *Proc. Inst. Food Sc. Technol.*, 10, 65–70. *Analyst*, 102, 225–68.

16. BRANCA, P. (1982). Contents of the oligoelements and chemical criteria for evaluating the age of canned goods (translation of title). *Boll. Chim. Unione Ital. Lab. Prov.*, 33, 495–506.

17. BIFFOLI, R., CHITI, F., MOCCHI, M. and PINZAUTI, S. (1980). Contamination of canned foods with metals (translation of title). *Rev. Soc. Ital. Sc. Alimentazione*, 9, 241–6.

18. HALLBERG, L. (1983). Prevention of iron deficiency in industrialised countries, in *Groupes à Risque de Carence en Fer dans les Pays Industrialises. Colloque Inserim-Ista/Cnam*, ed. DUPIN, H. and HERCBERG, S., 287–302 (Editions INSERM, Paris).

19. HALLBERG, L., BRUNE, M. and ROSSANDER, L. (1989). Iron absorption in man: ascorbic acid and dose-dependent inhibition by phytate. *Am. J. Clin. Nutr.*, 49, 140–4.

20. DISLER, P.B., LYNCH, S.R., CHARLTON, R.W. *et al.* (1975). The effects of tea on iron absorption. *Gut*, 16, 193–200.

21. LAYRISSE, M., MARTINES-TORRES, C., and GONZALES, M. (1974). Measurement of the total daily iron absorption by the extrinsic tag model. *Am. J. Clin. Nutr.*, 27, 152–62.

22. MONSEN, E.R. (1971). The need for iron fortification. *J. Nutr. Educ.*, 2, 152–5.

23. HALLBERG, L. (1982). Iron nutrition and food iron fortification. *Seminars in Haematol.*, 14, 31–41.
24. ROSSER, H.P. (1986). Iron. *J. Food Nutr.*, 42, 82–92.
25. POWELL, I.W., HALLIDAY, J.W. and BASSETT, M.C. (1982). The case against iron supplementation, in *The Biochemistry and Physiology of Iron*, ed. SALTMAN, P. and HEGENAUER, J., 811–43 (Elsevier, New York).
26. BULLEN, J.J., ROGERS, H.J. and LEIGH, L. (1972). Iron binding proteins in milk and resistance to *Escherichia coli* infection in infants. *Br. Med. J.*, I, 69–73.
27. BOTHWELL, T.H. and CHARLTON, R.W. (1982). A general approach to the problem of iron deficiency and iron overload in the population at large. *Seminars in Haem.*, 19, 54–62.
28. ALCARON, D.A., DONOVAN, M.E., FORBES, G.B. *et al.* (1979). Iron absorption in the thalassemia syndromes and its inhibition by tea. *New Engl. J. Med.*, 300, 5–8.
29. MACPHAIL, A.P., SIMON, M.O., TORRANCE, J.D. *et al.* (1979). Changing patterns of iron overload in black South Africans. *Am. J. Clin. Nutr.*, 32, 1272–8.
30. REILLY, C. (1973). Heavy metal contamination in home-produced beers and spirits. *Ecol. Food Nutr.*, 2, 43–47.
31. CHARLTON, R.W., HAWKINS, D.M., MAVOR, W.O. and BOTHWELL, T.H. (1970). Hepatic storage iron concentrations in different population groups. *Am. J. Clin. Nutr.*, 23, 358–65.
32. REILLY, C. (1985). The dietary significance of adventitious iron, zinc, copper and lead in domestically prepared food. *Food Add. Contam.*, 2, 209–15.
33. DROVER, D.P. and MADDOCKS, I. (1975). Iron content of native foods. *PNG Med. J.*, 18, 15–17.
34. WALKER, A.R.P. and ARVIDSSON, U.B. (1953). Iron 'overload' in the South African Bantu. *Trans. Roy. Soc. Trop. Med. Hyg.*, 47, 536–48.
35. BARR, D.B.G. and FRASER, D.K.B. (1968). Acute iron poisoning in children: role of chelating agents. *Br. Med. J.*, 1, 737.
36. AKATSUKA, K. and FAIRHALL, L.T. (1934). The toxicology of chromium. *J. Ind. Hyg. Toxicol.*, 16, 1–10.
37. SCHWARZ, K. and MERTZ, W. (1959). Chromium (III) and the glucose tolerance factor. *Arch. Biochem. Biophys.*, 85, 292–303.
38. UNDERWOOD, E. (1977). *Trace Elements in Human and Animal Nutrition*, 4th edn, 258 (Academic Press, New York and London).
39. BURROWS, D. (1983). *Chromium: Metabolism and Toxicity*, Preface (CRC Press, Boca Raton, Florida).
40. VEILLON, C. (1986). Trace element analysis of biological samples. Problems and precautions. *Anal. Chem.*, 58, 851A – 66A.
41. SMART, G.A. and SHERLOCK, J.C. (1985). Chromium in foods and the diet. *Food Add. Contam.*, 2, 139–47.
42. RASMUSSEN, G., (1983). *Release of Trace Elements from Kitchen Utensils.* Publication No. 77 (Miljöministerteriet, Denmark).
43. SANER, G. (1986). The metabolic significance of dietary chromium. *Nutr. Intn.*, 2, 213–20.
44. ANDERSON, R.A. and KOZLOVSKY, A.S. (1985). Chromium intake,

232   METAL CONTAMINATION OF FOOD

absorption and excretion of subjects consuming self-selected diets. *Am. J. Clin. Nutr.*, 41, 1177–83.
45. MURTHY, G.K., RHEA, U. and PEELER, J.R. (1971). Chromium in the diet of schoolchildren. *Environ. Sci. Technol.*, 5, 436–42.
46. MERANGER, J.C. and SMITH, D.C. (1972). The heavy metal content of a typical Canadian diet. *Can. J. Pub. Health*, 63, 53–7.
47. FOOD AND NUTRITION BOARD (1980). *Recommended Dietary Allowances* (National Academy of Sciences, Washington, DC).
48. BUNKER, V.W., LAWSON, M.S., DELVES, H.T. *et al.* (1984). The uptake and excretion of chromium by the elderly. *Am. J. Clin. Nutr.*, 39, 797–802.
49. OFFENBACHER, E.G. and PI-SUNYER, F.X. (1983). Temperature and pH effects on the release of chromium from stainless steel into water and fruit juices. *J. Ad. Food Chem.*, 31, 89–92.
50. JORHEM, L. and SLORACH, S. (1987). Lead, chromium, tin, iron and cadmium in foods in welded cans. *Food Add. Contam.*, 4, 309–16.
51. HOARE, W.E., HEDGES, E.S. and BARRY, B. (1965). *The Technology of Tinplate* (Edward Arnold, London).
52. WORLD HEALTH ORGANISATION (1970). *Standards for Drinking Water* (WHO, Geneva).
53. EUROPEAN ECONOMIC COMMUNITY (1980). *Council Directive Relating to the Quality of Water Intended for Human Consumption*, 80/778/EEC, 15 July.
54. DONALDSON, R.M. and BARRERAS, R.F. (1966). Intestinal absorption of trace quantities of chromium. *J. Lab. Clin. Med.*, 68, 484–9.
55. HILL, C.H. (1976). Mineral interrelationships, in *Trace Elements in Human Health and Disease*, ed. PRASAD, A.S., Vol. 2, 281–98 (Academic Press, New York).
56. MERTZ, W. (1969). Chromium occurrence and function in biological systems. *Physiol. Ref.*, 49, 163–239.
57. OKADA, S., TANIYAMA, M. and OHBA, H. (1982). Mode of enhancement in ribonucleic acid synthesis directed by chromium (III)-bound deoxyribonucleic acid. *J. Inorg. Biochem.*, 17, 41–9.
58. ROESCH, K., PATTERSON, C. and REILLY, C. (1985). Difficulties in the determination of total chromium in food. *Proc. Nutr. Soc. Aust.*, 10, 165.
59. ANDERSON, R.A., BOREL, J.S., POLANSKY, M.M. *et al.* (1988). Chromium intake and excretion of patients receiving total parenteral nutrition. Effects of supplemental chromium. *J. Trace Elem. Expt. Med.*, 1, 9–18.
60. WOLF, W., MERTZ, W. and MASIRONI, R. (1974). Determination of chromium in refined and unrefined sugars by oxygen plasma ashing flameless atomic absorption spectrophotometry. *J. Agric. Food Chem.*, 22, 1037–42.
61. CARY, E.E. and RUTZKE, E.E. (1983). Atomic absorption spectrophotometric determination of chromium in plants. *J. Assoc. Off. Anal. Chem.*, 58, 433–5.
62. VERSIECK, J. and CORNELIS, R. (1989). *Trace Elements in Human Plasma or Serum*, 68 (CRC Press, Boca Raton, Florida).
63. KUMPULAINEN, J. (1980). Chromium in foods and diets. *Mineral Elements '80: Proc.*, Part 1, 263–77.
64. WENLOCK, R.W., BUSS, D.H. and DIXON, E.J. (1979). Trace Nutrients 2. Manganese in British foods. *Br. J. Nutr.*, 41, 253–61.

65. DAVIDSON, L., CEDERBLAD, A., LONNERDAL, B. *et al.* (1989). Manganese retention in man: a method for estimating manganese absorption in man. *Am. J. Clin. Nutr.*, 49, 170–9.

66. SCHROEDER, H.A. (1971). Losses of vitamins and trace minerals resulting from processing and preservation of foods. *Am. J. Clin. Nutr.*, 24, 562–73.

67. WORLD HEALTH ORGANISATION (1980). *Recommended Health-Based Limits in Occupational Exposure to Heavy Metals. WHO Tech. Rep. Ser. No. 547* (WHO, Geneva).

68. CAWTE, J. and FLORENCE, M. (1987). Environmental source of manganese on Groote Eylandt, Northern Territory. *Lancet*, 1, 62.

69. CAWTE, J., HAMS, G. and KILBURN, C. (1987). Mechanisms in a neurological ethnic complex in Northern Australia. *Lancet*, 1, 61.

70. BELL, A. (1988). Tracking down the cause of a mysterious illness. *Ecos*, 57, 3–8.

71. CAWTE, J. and KILBURN, C. (1987). *Proceeedings of Conference, Darwin, Northern Territory, June* 1987 (Dept of Health, Darwin, N.T., Australia).

72. SCHROEDER, H.A., NASON, A.P. and TIPTON, I.H. (1967). Essential trace elements in man: cobalt. *J. Chron. Dis.*, 20, 869–90.

73. ANON. (1968). Epidemic cardiac failure in beer drinkers. *Nutr. Revs.*, 26, 173–5.

74. GRINVALSKY, H.T. and FITCH, D.M. (1969). A distinctive myocardiopathy occurring in Omaha, Nebraska: pathological aspects. *Ann. N.Y. Acad. Sci.*, 156, 544–65.

75. TOSKES, P.P., SMITH, G.W. and CONRADS, M.E. (1973). Cobalt absorption in sex-linked anaemic mice. *Am. J. Clin. Nutr.*, 26, 435–7.

76. YAMAGATA, N., MURATAN, S. and MORII, T. (1962). The cobalt content of the human body. *J. Radiat. Res.*, 3, 4–8.

77. BLOKHIMA, R.I. (1970). Cobalt and goitre, in *Trace Element Metabolism in Animals*, ed. MILLS, C.F. , 426–32 (Livingstone, Edinburgh).

78. ANDERSEN, I. and HEGETVEIT, A.C. (1984). Analysis of cobalt in plasma by electrothermal atomic absorption spectrometry. *Fresenius Z. Anal. Chem.*, 318, 41–8.

79. SMART, G.A. and SHERLOCK, J.C. (1987). Nickel in foods and diets. *Food Add. Contam.*, 4, 61–71.

80. MITCHELL, R.L. (1945). Trace elements in soil. *Soil Sci.*, 60, 63–75.

81. TSALEV, D.L. (1984). *Atomic Absorption Spectrophotometry in Occupational and Environmental Health Practice*, Vol. II, 152: Table 1: Nickel concentrations in human body fluids and tissues (CRC Press, Boca Raton, Florida).

82. MYRON, D.R., ZIMMERMAN, T.J., SHULER, T.R. *et al.* (1978). Intake of nickel and vanadium by humans. A survey of selected diets. *Am. J. Clin. Nutr.*, 31, 527–31.

83. NIELSEN, G.D. and FLYVHOLM, M. (1983). Risks of high nickel intakes with diets. *Symposium: Nickel in the Environment, Lyon, March* 8–11, 1983 (IARC/EEC/ILO).

84. ELLEN, G., VAN DEN BOSCH-TIBBESMA and DOUMA, F.F. (1987). Nickel content of various Dutch foodstuffs. *Z. Lebens. Unter- Forsch.*, 179, 145–56.

85. PENNINGTON, J.A.T. and JONES, J.W. (1987). Molybdenum, nickel, cobalt, vanadium, and strontium in total diets. *J. Am. Diet. Assoc.*, 87, 1644–50.

86. KOOPS, J., KLOMP, H. and WESTERBEEK, D. (1982). Spectroscopic determination of nickel with furildioxime with special reference to milk and milk products and to the release of nickel from stainless steel by acidic dairy products and by acid cleaning. *Netherlands Milk Dairy J.*, 36, 333–5.
87. BRUN, S., BLAISE, A., CABANIS, J.C. *et al.* (1983). Chemical migration of materials from wine containers (trans.). *Industrie delle Bevande*, 12, 345–53.
88. GESTRING, W.D. and JARRELL, W.M. (1982). Plant availability of phosphorus and heavy metals in soils amended with chemically treated sewage sludge. *J. Environ. Qual.*, 11, 669–75.
89. SUNDERMAN, F.W. (1965) Nickel toxicity. *Am. J. Clin. Path.*, 44, 182–200.
90. NIELSEN, F.H., MYRON, D.R., GIVAND, S.H. *et al.* (1975). Nickel deficiency in rats. *J. Nutr.*, 105, 1607–19.
91. STOKINGER, H.E. (1963). *Nickel in Industrial Hygiene and Toxicology*, Vol. II (Interscience, New York).
92. BROWN, S.S. and SUNDERMAN, F.W., Jr. (1980). *Nickel Toxicology* (Academic Press, London).
93. COUGHLAN, M.P. (1983). The role of molybdenum in human biology. *J. Inher. Metab. Dis.*, 6, *Suppl.* 1, 70–7.
94. RAJAGOPALAN, K.V. (1987). Molybdenum — an essential trace element. *Nutr. Revs.*, 45, 321–8.
95. TSONGAS, T.A., MEGLEN, R.R., WALRAVENS, P.A. *et al.* (1980). Molybdenum in the diet: an estimate of average daily intake in the United States. *Am. J. Clin. Nutr.*, 33, 1103–7.
96. BERROW, M.L. and WEBBER, J. (1972). Heavy metals in sewage sludge. *J. Sci. Food Agric.*, 33, 93–100.
97. POOLE, D.B.R. (1982). Bovine copper deficiency in Ireland — the clinical disease. *Irish Vet. J.*, 36, 169–73.
98. ANON. (1987). Molybdenum deficiency in TPN. *Nutr. Revs.*, 45, 337–41.
99. YAROVAYA, G.A. (1964). Molybdenum in the diet in Armenia. *Sixth Intn. Biochem. Cong., New York. Abstracts* 6, 440.

CHAPTER 10

THE OTHER TRANSITION METALS AND ZINC

The only other metals of the three transition series which are of any real significance as possible contaminants of food are titanium and vanadium of the first series, silver of the second, and tungsten of the third. Zinc, as has been observed earlier, is not strictly speaking a transition metal. However, it fits in chemically with the metals of the first series, completing the line from scandium to copper in the Periodic Table of the Elements. It shares with several of the biologically active transition metals common food sources, and interacts functionally with them in cells of the body. For these reasons, as well as for convenience, it will be treated here as an addendum to the transition metals.

TITANIUM

It may come as a surprise to many people to learn that titanium, a metal to which they seldom, if ever, have given any thought, is the eighth most common element in the earth's crust. Food scientists and nutritionists are generally no less ignorant about the metal than are members of the general public, for titanium does not normally warrant a place in their professional textbooks. However, as with several other ultra-trace components of the diet which are becoming of increasing commercial importance, it may be necessary to pay more attention to this element in the future. In fact, titanium is one of the very important 'strategic' metals and is used extensively in advanced engineering and space-age technology. It will be some time, however, before the large gap in our knowledge about the dietary role and metabolic significance of titanium is filled. At present we know very little about them.

PROPERTIES, PRODUCTION AND USES

Titanium is element number 22 in the Periodic Table, with an atomic weight of 47.9. It is a hard, light (density 4.5) metal, with a melting point of 1668° C. It is strongly resistant to corrosion. It can, however, be converted by oxygen into the commercially important titanium dioxide, TiO_2.

In spite of its abundance in the environment, titanium is difficult to extract. Major sources are mineral sands which contain the two important ores, ilmenite (ferrous titanate) and rutile (titanium dioxide). World production of titanium, which is rapidly growing, is about 1.5 million tonnes each year.

Titanium is used extensively in the aerospace and armaments industries, as well as in a variety of advanced engineering applications, including bioengineering technologies. Its lightness, combined with toughness and resistance to corrosion, make it ideal for such uses. A growing and very important use of the metal is for human and veterinary protheses. A minor, though important, use has already been referred to in this book in the section on sample preparation. Titanium corers and knives as well as homogeniser blades are used for the preparation of contaminant-free samples for trace element analyses of foods[1]. Titanium dioxide finds wide use as a white pigment in paints, enamels, paper coatings and plastics. In this use it has replaced the toxic white lead pigment. The oxide is also used in the food industry as an additive for whitening flour, confection and non-dairy milk product substitutes. These uses, and its applications in the pharmaceutical industry in a variety of drugs and cosmetic preparations, are a tribute to its perceived non-toxic nature.

TITANIUM IN FOOD AND BEVERAGES

Very little is known about levels of titanium in foods and the diet. Because of its wide-scale distribution in the environment, the metal is almost certainly a contaminant of all our foods and beverages, but in very low concentrations. This is mainly external contamination, since titanium is poorly absorbed by plants from the soil[2]. Levels of about 2.0 mg/kg have been reported in cereals, vegetables, dairy products and other foods. However, these reported concentrations are suspect, since techniques for the accurate determination of titanium at low levels in foods and biological tissues have not yet been well established. Inter-laboratory comparisons of analytical techniques and results have not been reported, certified reference materials for titanium are not available and reference values for the element in foods or biological tissues have not been established[3]. It is probable that actual levels of titanium in foods are 10–100 times lower than these reported values.

Some other data on levels of titanium in the diet and its components have been reported. Domestic water supplies contain about 2 μg/litre of the metal. Daily intake has been estimated to be between 3 and 600 μg in the US. It is possible that intake may be greater in the case of people living near coal- or oil-fired power stations[4].

ABSORPTION AND METABOLISM OF TITANIUM

Few studies have been reported on the uptake and behaviour of titanium in the human body. Very little of the element seems to be absorbed from ingested food and relatively high levels reported in lung tissue are believed to be due to dust inhalation[5]. Excretion of what is absorbed appears to be rapid. Titanium has no known physiological role and is not an essential nutrient. It is also reported to be essentially non-toxic, at least in the amounts normally consumed in food. As Nielsen[4] has noted, a review of titanium toxicity has questioned whether evidence of a specific oral toxicity has ever been found[6].

ANALYSIS OF FOODSTUFFS FOR TITANIUM

As has been observed above, little has been reported in the literature on analytical procedures for the determination of titanium in foods or biological tissues. Versieck and Cornelis recently commented that, as far as they were aware, 'the element titanium in human serum has hardly been searched for. That interest may, however, grow in order to learn the impact on the human body of the titanium material in modern prostheses and in many drugs and cosmetic preparations'[7].

Several instrumental methods of analysis for titanium can be used, depending on resources available. Graphite furnace atomic absorption, with deuterium lamp or Zeeman mode background correction, following acid digestion of samples, is reported to provide reliable results[8].

Neutron activation analysis has been used for the determination of titanium in serum by Cantone and colleagues[9].

VANADIUM

Vanadium has come to the attention of toxicologists and nutritionists over the past two decades, with recognition of its significance as an industrial toxin, and its essentiality to animals, probably including humans. There has been a considerable increase in studies of this once largely overlooked component of the diet. At present many questions about its role in human health and disease are still unanswered but, within another decade, many of the gaps in our knowledge of the element will be filled. As with some of the other 'ultratrace elements' that have been referred to already, some of the uncertainties about vanadium and its significance are due to the current absence of a reliable analytical technique for the element[10].

CHEMICAL AND PHYSICAL PROPERTIES

Vanadium is element number 23 in the Periodic Table, with an atomic number of 50.9. It is a relatively light metal (density 6.1), and is steel-grey in colour. It is tough and corrosion-resistant.

Vanadium has a number of oxidation states, from $+2$ to $+5$. The vanadate ion, VO_4^{3-} is the predominant form of the V(V) state in basic solution. The best known and stable form of the V(IV) state is the VO^{2+} ion. Vanadium is capable of undergoing oxidation–reduction exchanges and this suggests that at least some of its biological activities are related to cellular redox reactions.

PRODUCTION AND USES

Vanadium occurs very widely in the environment. It is present in soils at an average level of about 0.1 mg/kg[8]. It occurs in higher concentrations in ores, which are normally mixed. Carnotite, for instance, contains both uranium and vanadium. Vanadium is extracted as a by-product of uranium, as well as of iron and titanium metallurgy. An important source is fly ash from power stations, especially where oil is used as a fuel.

Much vanadium is used in the form of an iron alloy, ferrovanadium, which is easier to produce from mixed iron–vanadium ores than is pure vanadium. The alloy is used to give steel and cast iron shock resistance and ductility. Ferrovanadium is also combined with chromium and other metals for the manufacture of machines and tools requiring high-quality steels.

Pure vanadium and its salts have many other industrial uses. It serves as a catalyst in the production of ammonia, various polymers, and rubber. Its compounds are used in photography, printing and dyeing, as well as in the manufacture of pottery and glass[11].

VANADIUM IN FOOD AND BEVERAGES

Little reliable information is as yet available on normal levels of vanadium in individual foods and diets. Mean levels between 1 and 30 μg/kg have been reported in a variety of foods[12]. Levels in processed foods may be higher, up to 0.1 mg/kg. Pennington and Jones[13] give the following data on vanadium levels in certain food groups (expressed as means in μg/kg, with ranges in parentheses):

cereals/cereal products: 23 (0–150); meat/fish: 10 (0–120); vegetables: 6 (0–72); milk: 1 (0–6).

Some foods, especially seafoods, have been reported to have higher levels of the element. Vegetable oils have also been reported to contain high levels of vanadium, with some up to more than 40 mg/kg[14].

Estimated daily intakes range from 10 to 30 μg. A UK Total Diet Study found an intake of 13.0 μg/day in adults[15], but values of up to 230 μg have been reported from Japan[8]. As Pennington and Jones have noted[13], earlier reported intakes of vanadium were considerably higher (up to 2 mg/day) than those published since the late 1970s.

ABSORPTION AND METABOLISM OF VANADIUM

It is believed that ingested vanadium is only poorly absorbed, with between 0.1 and 2 per cent entering the bloodstream from the gut. Daily absorption has been estimated to be between 5 and 10 μg[16]. Vanadium is found in extremely low concentrations in biological tissues, including blood. Excretion of accumulated vanadium from the body is mainly in urine.

Vanadium has been shown to be an essential trace nutrient for a variety of animals. Vanadium deficiency is associated with stunted growth, impaired reproduction, altered red blood cell formation and iron metabolism, changes in blood lipid levels and hard tissue formation. There is a growing belief among health experts that the metal can play similar roles in humans. Vanadium has been shown to be capable of modifying the activity of a number of enzymes, including Na-K-ATPase which is important in muscle contraction, and tyrosine kinase which is located in growth factors, oncogenes, phosphatases and receptors for insulin[17].

Dietary vanadium does not appear to be toxic to man, at least at the low levels normally encountered in foods. However, it has been shown to be toxic to rats at relatively low concentrations. Industrial exposure of workers to vanadium dusts is a well-recognised occupational hazard. It results in eye and lung irritation, as well as inhibition of action of the enzyme cholinesterase which results in a deficiency of choline. Defects in metabolism of the sulphur-containing amino acid cysteine also occur. Other effects, including dermatitis and anorexia, have also been reported[8].

ANALYSIS OF FOODSTUFFS FOR VANADIUM

As has been mentioned, there are few reports in the literature of reliable methods for the analysis of vanadium in foods. Very low levels of concentrations and the possibility of contamination from the environment are a particular problem.

Graphite furnace atomic absorption spectrophotometry may be used, following preconcentration by solvent extraction or use of ion exchange techniques[7]. However, problems of low sensitivity are difficult to overcome.

Iyengar and Wottiez[3] recommend that because of the problems of contamination, neutron activation analysis, preferably without pre-irradiation chemical separation, be used since it offers the best chance of obtaining accurate results.

Berner and his colleagues, in their study of ultratrace elements in total parenteral nutrition fluids[17], used inductively coupled argon plasma emission spectrometry. This technique gave accurate results, at low levels of concentration, with the added advantage of simultaneous determination of a variety of trace elements.

SILVER

CHEMICAL AND PHYSICAL PROPERTIES

Silver is element number 47 in the Periodic Table, with an atomic weight of 107.87. It is a white, lustrous, soft and malleable metal, capable of taking a high degree of polish. Its melting point is 960° C. It has the highest known electrical and thermal conductivities of any metal, properties which account for its use in the electrical industry as well as in the traditional kitchen.

Silver is relatively unreactive chemically, but it blackens in the presence of sulphur and hydrogen sulphide as silver sulphide is formed. Several oxidation states are known, though only Ag(I) is of practical consequence. Argentous compounds of importance are Ag(I) nitrate, acetate and its halides.

PRODUCTION AND USES

Silver has long been valued as a precious metal, used for the manufacture of ornaments, as well as for the kitchen and boudoir utensils of the rich. Since Roman times it has been extracted as a by-product of lead production by cupellation of silver-containing lead ores. It is also extracted from other mixed ores, as well as from the silver ore, argenite. Silver is produced on a large scale in Canada, Peru, the USSR, USA and Mexico. Smaller deposits are worked, often on a family scale, in many parts of the world.

Apart from its use in jewellery and coins, silver is used in tableware. It forms a number of useful alloys and solders with copper, cadmium and lead. Its compounds are widely used in the photographic industry, as a germicidal agent in water treatment and for other related purposes. There are a number of pharmaceutical applications of the metal and its compounds.

SILVER IN FOOD AND BEVERAGES

No extensive studies on levels of silver in foods have been reported. Food, in fact, seems to contain very little silver. Reported levels range from undetectable to a few micrograms per kilogram of fruits and vegetables and little more in other foods[4]. Intake is probably about 20 μg/day[18].

Where silver-plated cooking utensils are used to prepare food, intake of the metal may be increased slightly. The use of silver and its compounds to treat domestic water supplies may also contribute an additional amount of the metal to the diet. There is some evidence that certain vegetables, such as some varieties of *Brassica*, may accumulate higher than usual levels of silver, especially if grown close to coal-burning power stations, but the level of increase is unlikely to cause an excessive dietary intake of the metal[19].

ABSORPTION AND METABOLISM OF SILVER

Little is known about the level of absorption of silver from ingested food. Evidence from animal studies suggests that it may be as low as 10 per cent. The level of the metal in normal tissue is also low. When excessive intake occurs, accumulation appears to take place in liver as well as in skin. Most of the silver that is absorbed is rapidly excreted, mainly in bile with faeces and a small amount in urine.

Silver is not shown to be essential for animals, including man. It has been shown to interact metabolically with copper and selenium. In experimental animals silver accentuates signs of copper deficiency and induces a selenium-like deficiency state. Silver nitrate or silver lactate in drinking water was found to promote necrotic liver degeneration in vitamin E-deficient rats[4]. The addition of both copper and selenium to the diet has been reported to decrease the toxicity of silver to turkeys[20].

While there is no evidence that silver in normal diets has a toxic effect on humans, prolonged ingestion of silver in pharmaceutical preparations can bring about the development of a distinctive blue-grey discolouration of the skin, eyes and mucous membranes. This condition, known as *argyria*, is also seen sometimes in silversmiths and others who are industrially exposed to the metal[21]. The consumption of certain 'anti-smoking lozenges' which contain silver salts has been reported to bring about a similar effect. There is no evidence that argyria is caused by levels of silver normally found in the diet.

ANALYSIS OF FOODSTUFFS FOR SILVER

Because of the low concentrations of silver in foods and biological materials, problems due to inadequate sensitivity occur with several analytical procedures. Sample preparation is critical. Careful wet digestion of samples, using a nitric–sulphuric–perchloric acids mixture is advisable to prevent losses through volatilisation. Dry ashing is unacceptable since silver is readily reduced to the metallic state and volatilises at ashing temperatures.

Electrothermal atomic absorption spectrophotometry is satisfactory for end determination. Background correction is always necessary, using a deuterium lamp or, preferably, Zeeman mode correction techniques.

Chelate extraction with sulphur-containing ligands, such as dithiocarbamates or dithiozone, is suitable for very low concentrations of silver in biological samples[8].

TUNGSTEN

CHEMICAL AND PHYSICAL PROPERTIES

Tungsten is element number 74 in the Periodic Table, with an atomic weight of 183.5. It has the chemical symbol W, from its alternative name wolfram. It is a heavy metal, with a density of 19.3. Its melting point is 3410° C, which is the highest melting point of any metal. Tungsten is a grey, hard, ductile, malleable metal, highly resistant to corrosion.

Tungsten is a transition metal of the third series, and like the other metals of the group, it has several oxidation states. It is not a very reactive metal. It forms an industrially important oxide, tungsten trioxide, WO_3 and tungstic acid, hydrated trioxide, H_2WO_4.

PRODUCTION AND USES

The main ores of tungsten are wolframite, $FeWO_4$, and scheelite, $CaWO_4$. China is a major source of tungsten ores. The metal is extracted by converting the ore to the oxide by roasting and then reducing the oxide with hydrogen to the free metal.

Tungsten has many important industrial uses. It is mainly used in high-speed steels and special alloy steels. Tungsten increases the hardness of steel, making it ideal for tools, and for forming dies for die-casting. Tools such as chisels and punches made from tungsten are strong and resistant. They have a close uniform texture and can be ground to a fine edge.

A compound made from tungsten and carbon, called tungsten carbide, forms the hardest metallic substance known. It is known as Carboloy and is next to the diamond in hardness. It is used to make cutting surfaces on bits for drills and machining tools.

Tungsten is used as a filament in electric light bulbs because of its high melting point and strength. Compounds of the metal are used as pigments.

TUNGSTEN IN FOODS AND DIETS

Little is known about levels of tungsten in foods. They appear to be very low, in spite of the wide industrial use of the metal. Daily intake in Sweden has been estimated to be 8–13 μg, with levels in domestic water supplies between 0.03 and 0.1 μg/litre[22]. A UK study found daily intake of the metal to be less than 1 μg[23].

ABSORPTION AND METABOLISM OF TUNGSTEN

Absorption of tungsten from food appears to be very limited. In experimental animals 97 per cent of an ingested dose of the metal was eliminated from the body in 72 h. Excretion was found to be equally divided between urine and faeces[4]. There may be some retention of the element in bones.

Tungsten is virtually innocuous to experimental animals and, presumably, to man. There is no evidence that it is an essential nutrient for any animal. However, animals fed a tungsten-containing feed were found to have reduced activities of the two molybdenum-containing enzymes xanthine oxidase and sulphite oxidase. Whether a similar antagonism occurs in humans is not known.

ANALYSIS OF FOODSTUFFS FOR TUNGSTEN

The low levels of tungsten in foods and the absence of well-established analytical techniques for the metal in biological samples present problems for the food analyst interested in the metal. Inorganic solutions of tungsten can be analysed readily using graphite furnace atomic absorption spectrophotometry. Zeeman mode or deuterium lamp background correction will have to be applied when samples with very low concentrations of the element are being investigated.

ZINC

Zinc merits a special place in a book dealing with metals in food. It is probably one of the most thoroughly investigated of all the nutritionally important trace elements. The recognition in the early 1960s of a nutritional need for the metal, and of the existence of clear deficiency states in certain population groups, helped trigger an interest in trace elements among life scientists that has not ceased to expand in the past three decades. Yet, in spite of its significance to food science and nutrition, and its prominence in the scientific literature, zinc has been described as the 'unassuming nutrient'[24].

There are good reasons for describing zinc in such terms. In spite of its important role in our biological and economic lives, it has never, outside of scientific circles, attracted the attention that lead and mercury, and chromium and selenium, for example, have. Even among scientists, though literature citations are still numerous, it is being replaced as a topic of research by several of the newer 'ultratrace elements' such as aluminium.

It comes as a surprise to many to learn that the human body contains almost as much zinc as it does iron. In fact an adult male normally contains 2–3 g of zinc, 3–4 g of iron and only about 10 mg of copper, the next most plentiful essential trace element.

More than 200 enzymes in the living world are known to contain zinc and more than 50 of these have important metabolic roles in animals. Zinc is an essential nutrient, no less important to the body than protein or ascorbic acid, or any other essential component of the diet. In the absence of an adequate dietary supply of zinc, serious and characteristic deficiency symptoms arise. Yet this metal has been overlooked as a contributor to the functions of human life until relatively recently.

To some extent the failure by biochemists and other life scientists to recognise the importance of zinc was due to its physico-chemical properties. Unlike its neighbouring transition elements, zinc does not form characteristic, brightly coloured complexes with reagents. This made it less easy to detect in foods and biological tissues than, say, copper or iron. Neither does zinc, though it is an industrially important metal, play a major role in construction or manufacture in its own right. It is, rather, the handmaiden of iron and other metals to which it lends protection in galvanising or by forming an alloy. Zinc, unlike iron or the copper alloy bronze, never had the distinction of having an age of human history named after it, in spite of its importance to human economic development.

Today, however, there is recognition among many scientists and health professionals of the major importance of zinc in human health. At the same time there is a steady increase in world production and use of the metal. The supreme accolade, in nutrition circles, was paid to the metal when, in 1978, it was included for the first time in the tables of food composition in the UK 'nutritionist's bible', *McCance and Widdowson's Composition of Foods*[25]. Zinc, since that date, has been included in food composition tables published in most developed countries.

CHEMICAL AND PHYSICAL PROPERTIES

Zinc is element number 30 in the Periodic Table, with an atomic weight of 65.37. Its density is 7.14, which makes it one of the 'heavy' metals. It is a bluish-white, lustrous metal, with a low melting point of 420° C. It is ductile and malleable when heated to only 100° C. Zinc tarnishes in air to a blue-grey colour as a coat of basic zinc carbonate, $Zn_2(OH)_2CO_3$, forms. This adhering layer protects the underlying metal from further corrosion. It is a tough, enduring coat and is the reason why zinc is used so extensively, in galvanising, on other less resistant metals.

Zinc is a reactive metal. It combines readily with non-oxidising acids, releasing hydrogen and forming salts. It also dissolves in strong acids to form zincate ions, $(ZnO_2)^{2-}$. The metal reacts with oxygen, especially at higher temperatures, to form zinc oxide. It also reacts directly with halogens and with sulphur and other non-metals. In spite of its completely filled $3d$ electron shell, zinc shares with its neighbouring transition metals an ability to form strong complexes with organic ligands.

Zinc forms some commercially very important alloys. The best known of these is brass. This alloy has been produced for more than 2000 years and is still widely used. A variety of zinc-containing bronzes are also produced.

PRODUCTION AND USES

Zinc ores are widely distributed and mines are operated in several European countries as well as on a larger scale in the USA, USSR, southern Africa and Australia. One of the largest deposits commercially exploited in Europe is in Ireland, with other large mines in Yugoslavia and Spain. The principal ores are sulphides, such as zinc blende. The carbonate calamine, oxide zincite, and silicate willemite, are also worked commercially. Zinc normally occurs in association with other metals, such as lead, cadmium and copper and the metals are extracted in a combined operation.

Zinc ore is crushed and may be concentrated by a flotation process, before it is roasted to produce the oxide. This is then reduced by heating with coke and the metal is distilled off. Zinc, which is very volatile, can produce large emissions during smelting and environmental pollution can be a serious problem.

In 1988 the International Lead and Zinc Study Group estimated that consumption of zinc in the Western world increased by about 4 per cent from the previous year to 5.2 million tonnes per annum[26]. Not all of this metal is from primary production since recovery from scrap accounts for about one-sixth of the total used.

The main use of zinc for many centuries was in the production of brass. This alloy was produced in large quantities in China and India and was an important item of trade when European merchantmen began to penetrate the Pacific Ocean. Today brass is still of importance and continues to be used, as it was traditionally, for the manufacture of cooking equipment and food and beverage containers. However, this use has declined considerably in the West following the introduction of aluminium and other metals for making pots and pans. Visual evidence from the markets of many Eastern countries suggests that the same is happening there also.

A major use of zinc is to protect iron and other metals from corrosion from air and water. In a process called galvanising, a coat of zinc is applied to other metals by various methods such as hot dipping or by dusting (sherardising). The external layer of zinc soon undergoes corrosion to form a protective skin of basic bicarbonate. The zinc also protects the other metals by acting as a 'sacrificial' metal. As it corrodes, the zinc releases electrons which flow to the other metal making it more negative in potential and thus reducing its corrosion rate. The electrical circuit, from the zinc anode to the iron or steel cathode, is completed when the electrons return through water or another conducting medium to the zinc. Sacrificial zinc anodes are also used to protect ships' hulls, offshore oil production platforms, pipelines and other iron and steel structures exposed to seawater.

About half of the world's zinc is used in galvanising, much of it in the automobile industry. About one-third of zinc produced in the US is used in this way, with much of the remainder employed for a similar purpose in the construction industry.

Zinc has many applications in the chemical and pharmaceutical industry. A considerable quantity is used in the manufacture of dry cell batteries. It is also used in the manufacture of paints and other pigments as well as rubber and plastics. Zinc is used in the manufacture of cosmetics, medicines, nutritional supplements and a variety of other pharmaceutical and household consumables. Its wide use in the home and elsewhere in the community accounts for a very high level of zinc contamination encountered in dust, water and the atmosphere, a hazard well known to the analyst[27].

ZINC IN FOOD AND BEVERAGES

Zinc occurs in all foods and beverages, in amounts related to the type of food as well as to the amount of exposure and types of processing to which it has been subjected. Numerous databases now exist for levels of the metals in very many types of foods, and in the diets of different groups of people. While there are considerable differences between levels of zinc in different kinds of food, generally data published in different countries are comparable for the same kinds of foods. Recent studies in the author's laboratory have provided the following data for different types of foods collected, as eaten, in a dietary survey (expressed as means, in mg/kg, with ranges in parentheses):

meat/fish: 35.5 (5.1–86.00); cereal/cereal products: 9.5 (0.7–19.0); vegetables: 2.5 (0.2–4.5); milk: whole, cow's 4.3 (3.8–4.7); cheese: cheddar, 38.2 (12.4–47.6).

Higher levels of zinc were found in certain commodities. Multigrain wholemeal bread contained between 8.0 and 13.4 mg/kg; bran breakfast cereals, unprocessed, had 49 mg/kg, and frozen peas, up to 23 mg/kg of the metal.

These results are comparable to those published in the UK[25], the US[28] and for the Australian Market Basket Survey[29]. As the latter paper notes, foods that make a significant contribution to the zinc content of diets are meats, especially organ meats, wholegrain cereals and milk products, including cheese. In addition, for those who consume them, oysters, with more than 100 mg/kg and certain nuts (peanuts had nearly 30 mg/kg), may make a significant contribution to intake of zinc.

An interesting comment on how food choice can affect intakes of zinc, and other metals which are not evenly distributed between all types of food, is made in a study on trends in zinc intakes in the US since early in the century[30]. The authors note that early in the century zinc was provided in almost equal proportions by foods of animal and vegetable origin, whereas in recent decades animal products have provided approximately 70 per cent of intake. They found that while three food groups — meat/poultry/fish, dairy products, grain products — account for 75–80 per cent of zinc intake in the US, over the years the proportion contributed by cereals has declined, while that contributed especially by the meat/poultry/fish group has increased.

Daily intake of zinc by adults in the UK[31] has been calculated to be 13.0 mg and in the US[32] between 10 and 15 mg. Estimates of daily intake in the Netherlands are from 15.4 to 24.6 mg[33]. A recent Australian study found that intake by a group of healthy children ranged from 5.0 to 11.8 mg/day, with a mean of 7.0[34].

ABSORPTION AND METABOLISM OF ZINC

The gastro-intestinal absorption of zinc is affected by a number of factors and is highly variable. Uptake has been reported to range from less than 10 to more than 90 per cent, with an average of about 20–30 per cent[35]. Various components of the diet can alter intake from food. Phytate, fibre and calcium can restrict absorption, whereas animal protein enhances uptake. A diet rich in wholemeal bread, which is rich in all three antagonists of uptake, can result in a deficiency of the element. This is particularly so if the bread is unleavened. If yeast is present during the fermentation of dough, some of the phytate is broken down by enzymes. Widespread chronic zinc deficiency occurs in parts of Iran and elsewhere in the Middle East as the result of consumption of such a diet. This resulted in the zinc deficiency of an epidemic nature, reported by Prasad and his colleagues in the early 1960s. Their discovery was the first indication that zinc deficiency in humans could be a major health problem[36].

There is considerable evidence of competition between zinc and other metals for uptake. Competitive interactions between zinc and copper, iron, cadmium and tin have been shown in animals. In humans high zinc intake can reduce copper absorption. Iron to zinc ratios of 2:1 or greater can reduce zinc absorption. Competition with tin has also been reported to occur in humans[37].

As well as animal protein, it is believed that certain small ligands in foods facilitate uptake of zinc from the diet. Human, but not cow's milk, may contain such a ligand, which it has been suggested accounts for the greater availability of zinc from breast milk than from bottle feeds.

Absorbed zinc is bound to albumin or transferrin in the blood. There is evidence that absorption of zinc is related to the level of metallothionein in blood. It has been suggested that a reliable method for assessing zinc nutritional status in the body, a task for which at present there is no reliable procedure, could be based on measurement of metallothionein levels in blood or urine[38]. Zinc is rapidly accumulated in organs and tissues, especially the liver, pancreas, spleen and kidney. Highest concentrations are found in the prostate gland and the eye. Skin, nails, hair, muscle and blood also contain appreciable amounts of zinc. Levels in hair have been used to determine dietary uptake of the metal[39]. There is controversy about the accuracy of such assessments.

Zinc is rapidly excreted via the faeces. These contain both unabsorbed zinc as well as that which had been absorbed and then secreted into the intestine, primarily by the pancreas. Only a small amount of zinc appears in the urine, normally about 0.5 mg/day. The same amount may be excreted in sweat. This loss may be greatly increased when excessive sweating takes place.

BIOLOGICAL EFFECTS OF ZINC

Zinc is now well established as a dietary essential for humans. Recommended dietary intakes have been established in many countries and the importance of an adequate intake at every stage of life is universally accepted. Several distinctive diseases related to zinc deficiency, from acrodermatitis enteropathica to alcohol-induced hyperzincuria, are recognised and are successfully treated with zinc supplements.

The biochemical bases of the metabolic roles of zinc are fairly well understood and our understanding of the function of the metal in the human body continues to expand. In man zinc enzymes occur in each of the six categories designated by the International Union of Biochemistry's Commission on Enzyme Nomenclature. These include nucleic acid polymerases, lactic acid, alcohol, and retinol dehydrogenases, phosphatases, proteases and several others such as carbonic anhydrase. As Mills has pointed out[40], zinc, as well as being a component of enzymes, is also part of many non-enzyme proteins. Indeed, non-enzyme zinc makes up the bulk of the body's zinc content and is 'not a trivial adventitious contaminant'. It has an important role in growth, in preventing pathology, in modifying protein structure and many other functions. Zinc, according to Mills, has more known functions in the body than any other inorganic nutrient. Among these is its role in the beta-cells of the pancreas where it appears to stabilise the insulin molecule. It functions, possibly in a similar manner, in the choroid region of the eye in binding the retina in position.

The multiple roles of zinc are indicated by what occurs when zinc is deficient in the diet of young animals. There is a rapid development of clinical symptoms which include growth retardation, poor and erratic consumption of food, loss of appetite, alopecia, development of scaly keratinous lesions on the skin. Resistance to infection is reduced through failure of the body's immune response mechanism. Wound healing is slowed. Conditions corresponding to these have been observed, also, in humans. An extreme example is the inherited disease acrodermatitis enteropathica, in which there is an inability to absorb an adequate amount of zinc from the diet. It is characterised by severe bullous-pustular dermatitis of the extremities and the oral, anal and genital areas, and total hair loss. Similar symptoms, of equal or lesser severity, are sometimes seen in iatrogenic or conditioned zinc deficiency, due to side-effects of treatment with such chelating agents as penicillamine, used in Wilson's disease, or to a high fibre and phytate diet such as was consumed by the Iranian victims of zinc deficiency. Stunted growth, dwarfism, failure to mature sexually and alopecia, are symptoms of such deficiency. Such deficiency can result in death. In all these cases therapy with zinc supplements is beneficial and will normally bring about a restoration of normal health.

TOXIC EFFECTS OF ZINC IN FOOD AND BEVERAGES

Zinc salts are intestinal irritants and can at a certain level of intake, usually 250 mg or more, cause nausea, vomiting and abdominal pain. However, often zinc is accompanied by cadmium, and this, not the zinc, may be the cause of the intestinal distress.

There are a number of reports in the literature of incidents of poisoning from the use of galvanised iron containers to hold acidic drinks, such as orange juice or alcoholic beverages.

Longer term effects of high zinc intakes have been reported by several investigators[41]. These were related to interference by zinc with copper absorption and were, in effect, symptoms of copper deficiency. It is suggested that a high zinc to copper ratio may result in interference with lipid metabolism and cause hypercholesterolemia[42].

ANALYSIS OF FOODSTUFFS FOR ZINC

Zinc is readily determined in food by atomic absorption spectro-photometry. Depending on levels in samples, flame or graphite furnace AAS may be used, with good results.

In the author's laboratory food samples are digested in a nitric–sulphuric–perchloric acids mixture and are analysed by flame or Zeeman background corrected AAS, with good recovery and adequate sensitivity. There are no major problems from interference. However, it is important that great care be taken to prevent contamination of samples. Zinc is a pervasive and universally distributed component of the air and dust, and occurs in many of the materials used in laboratories. Good hygiene practices can overcome this problem.

REFERENCES

1. KUMPULAINEN, J. and PAAKKI, M. (1987). Analytical quality control program used by the trace elements in foods and diets sub-network of the FAO European cooperative net-work on trace elements. *Fresenius Z. Anal. Chem.*, 326, 684–9.
2. MITCHELL, R.L. (1957). Mineral composition of grasses and clovers grown on different soils. *Research (London)*, 10, 357–62.
3. IYENGAR, V. and WOTTIEZ, J. (1988). Trace elements in human clinical specimens: evaluation of literature data to identify reference values. *Clin. Chem.*, 34, 474–81.
4. NIELSEN, F.H. (1986). Other elements, in *Trace Elements in Human and Animal Nutrition*, 5th edn, ed. MERTZ, W., Vol. 2, 415–62 (Academic Press, New York & London).
5. TIPTON, I.H. and COOK, M.J. (1963). Titanium in human tissues. *Health Phys.*, 9, 103–45.
6. ANON. (1980). In *Mineral Tolerance of Domestic Animals* (National Academy of Sciences, Washington, DC) [reference from Nielsen (1986)].
7. VERSIECK, J. and CORNELIS, R. (1989). *Trace Elements in Human Plasma or Serum* (CRC Press, Boca Raton, Florida); LUGOWSKI, S., SMITH, D.C. and VAN LOON. J.C. (1987). The determination of titanium and vanadium in whole blood, *Trace Elem. Med.*, 4, 28–34.
8. TSALEV, D.L. (1984). *Atomic Absorption Spectrometry in Occupational and Environmental Health Practice* (CRC Press, Boca Raton, Florida).
9. CANTONE, M.C., MOLHO, N. and PIROLA, L. (1985). Cadmium and titanium in human serum determined by proton nuclear activation. *J. Radioanal. Nucl. Chem.*, 91, 197–83.

10. WORLD HEALTH ORGANISATION (1988). *Vanadium. Environmental Health Criteria,* No. 81 (United Nations Environment Programme, International Labor Organisation, World Health Organisation, Geneva).
11. BERMAN, E. (1980). *Toxic Metals and their Analysis* (Heyden, London).
12. MYRON, D.R., GIVAND, S.H. and NIELSEN, F.H. (1977). Vanadium content of selected foods as determined by flameless atomic absorption spectroscopy. *J. Agric. Food Chem.,* 25, 297–300.
13. PENNINGTON, J.A.T. and JONES, J.W. (1987). Molybdenum, nickel, cobalt, vanadium, and strontium in total diets. *J. Am. Diet. Assoc.,* 87, 1644–50.
14. SCHROEDER, H.A. and NASON, A.P. (1971). Trace element analysis in clinical chemistry. *Clin. Chem.,* 17, 461–74.
15. EVANS, W.H., READ, J.I. and CAUGHLIN, D. (1985). Quantification of results for estimating elemental dietary intakes of lithium, rubidium, strontium, molybdenum, vanadium and silver. *Analyst,* 110, 873–86.
16. MYRON, D.R., ZIMMERMAN, T.J., SHULER, T.R. *et al.* (1978). Intake of nickel and vanadium by humans: a survey of selected diets. *Am. J. Clin. Nutr.,* 31, 527–31.
17. BERNER, Y.N., SHULER, T. R., NIELSEN, F.H. *et al.* (1989). Selected ultratrace elements in total parenteral nutrition solutions. *Am. J. Clin. Nutr.,* 50, 1079–83.
18. HARVEY, S.C. (1980). In *The Pharmacological Basis of Therapeutics,* eds GOODMAN, L.S. and GILLMAN, A., 967–9 (Macmillan, New York).
19. HEADLEE, A.J.W. and HUNTER, R.G. (1953). Uptake of metals by vegetables in the vicinity of coal-fired power stations. *Ind. Eng. Chem.,* 45, 548–51.
20. JENSEN, S.L. (1974). Trace elements in the diet of turkeys. *Poultry Sci.,* 53, 57–64,
21. HILL, W.B. and PILLSBURY, D.M. (1939). *Argyria. The Pharmacology of Silver* (Williams, Baltimore).
22. WESTER, P.O. (1974). Tungsten in Swedish diets. *Atheroscler.,* 20, 207–15.
23. HAMILTON, E.I. and MINSKI, M.J. (1972). Tungsten in the UK diet. *Sci. Total Environ.,* 1, 375–94.
24. REILLY, C. (1978). Zinc, the unassuming nutrient, in *Getting the Most out of Food,* Vol. 13, 47–69 (Van den Bergh and Jurgens, London).
25. PAUL. A.A. and SOUTHGATE, D.A.T. (1978). *McCance and Widdowson's The Composition of Foods* (HMSO–Elsevier, London and Amsterdam).
26. ANON. (1988). Elements, lead and zinc in the market place, *Metals Review,* No. 16, October (Australian Lead and Zinc Development Associations, Melbourne).
27. KOSTA, L. (1982). Contamination as a limiting parameter in trace analysis. *Talenta,* 29, 985–92.
28. MURPHY, E.W., WILLIS, B.W. and WATTS, B.K. (1975). Provisional tables on the zinc content of foods. *J. Am. Diet. Assoc.,* 66, 345–55.
29. ENGLISH, R. (1981). The Market Basket (Noxious Substances) Survey: zinc content of Australian diets. *J. Food Nutr.,* 38, 63–5.
30. WELSH, S. O. and MARSTON, R.M. (1983). Trends in levels of zinc in the US food supply, 1909–1981, in *Nutritional Bioavailability of Zinc,* ed. INGLETT, G.E., 15–30 (American Chemical Society, New York).

31. MINISTRY OF AGRICULTURE, FISHERIES AND FOOD (1976). *Manual of Nutrition*, 8th edn (HMSO, London).
32. FOOD AND NUTRITION BOARD (1980). *Recommended Dietary Allowances*, 9th edn (National Academy of Sciences, Washington, DC).
33. REITH, J.F., ENGELSMA, J.W. and VAN DITMARSH, W.C. (1976). Zinc in diets in the Netherlands (trans). *Voeding*, 37, 498–507.
34. BARRETT, J., REILLY, C., PATTERSON C. *et al.* (1990). Trace element nutritional status and dietary intake of children with phenylketonuria. *Am. J. Clin. Nutr.*, 52, 159–65.
35. FORBES, R.M. and ERDMAN, J.W.Q. (1983). Bioavailability of trace mineral elements. *Ann. Rev. Nutr.*, 3, 213–31.
36. PRASAD, A.S. (1983). Human zinc deficiency, in *Biological Aspects of Metals and Metal-Related Diseases*, ed. SARKAR, B., 107–19 (Raven Press, New York).
37. SOLOMONS, N. W. (1983). Competitive mineral–mineral interaction in the intestine, in *Nutritional Bioavailability of Zinc*, ed. INGLETT, G.E., 247–71 (American Chemical Society, New York).
38. BREMNER. I. and MORRISON, J.N. (1988). Metallothionein as an indicator of zinc status, in *Essential and Toxic Trace Elements in Human Health and Disease*, 365–79 (A. Liss, New York).
39. REILLY, C. and HARRISON, F. (1979). Zinc, iron, copper and lead in scalp hair of students and non-students in Oxford. *J. Hum. Nutr.*, 33, 250–4.
40. MILLS, C.F. (1988). The biological significance of zinc in man: problems and prospects, in *Zinc in Human Biology*, ed. MILLS. C.F., Chap. 24 (Springer, Berlin).
41. FOX, M.R.S. and JACOBS, R.M. (1985). Human nutrition and metal ion toxicity, in *Metal Ions in Biological Systems*, ed. SIGEL, H., 201–28 (M. Dekker, New York and Basel).
42. KLEVAY, L.M. (1975). The relation of zinc and copper to blood lipid levels. *Am. J. Clin. Nutr.*, 28, 764–72.

CHAPTER 11

BERYLLIUM, STRONTIUM, BARIUM AND THE OTHER METALS — SUMMING UP

The 20 metals that have so far been considered, in varying detail, in this book make up only a quarter of the elements that come under the definition of metal or metalloid. The other 60, while many are of considerable commercial and chemical interest, are not considered to be of major consequence as contaminants of food. However, that situation may change for, under certain circumstances, these other elements could also occur in food. The significance of such contamination for the consumer, as well as for the manufacturer, may in some cases be considerable. With increasing use of what were once 'rare earths' and other 'space-age' elements, the likelihood of their contaminating the environment, and with it food, is constantly increasing. It will be useful, therefore, to look briefly at some of these other potential contaminants from the point of view of present uses and in anticipation of future developments.

BERYLLIUM

Beryllium might easily have been moved to an earlier chapter as concern about its toxicity and expanding use in industry grows in the community. The reassuring press release by the US Public Health Service in 1943[1], that beryllium was not harmful, has long since been recognised as a serious and regrettable error.

CHEMICAL AND PHYSICAL PROPERTIES

Beryllium belongs, along with magnesium, calcium, barium and strontium, to Group IIA of the Periodic Table. These metals are known as the

255

alkaline earth metals. Beryllium is element number 4 in the Table, with an atomic weight of 9.0. It is a light metal, with a density of 1.84. It is silvery white in colour and very strong and flexible, with a high melting point. It is corrosion resistant, transparent to electrons, reflects neutrons and has low coefficient of expansion.

Chemically, beryllium resembles aluminium. Its salts are readily hydrolysed. In water the hydroxide, $Be(OH)_2$, is amphoteric. When it is dissolved in alkali, compounds of the type Na_2BeO_2 are formed.

PRODUCTION AND USES

Beryllium occurs in a number of ores, best known of which are beryl (beryllium aluminium silicate) and bertrandite (hydrated beryllium disilicate). The metal is extracted from the ore by electrolysis.

The physical properties of beryllium make it ideal for a number of very important industrial uses: as windows for X-ray tubes, in equipment requiring high electrical and thermal conductivity, and in springs that must resist frequent vibrations. Beryllium also forms a number of valuable alloys, with copper and other metals. These are used in making non-sparking tools, dies, high-strength, lightweight parts for aircraft, missiles and nuclear reactors. Beryllium is a metal well suited to the nuclear and space age, and its applications are constantly increasing in number.

Beryllium oxide was formerly used in fluorescent tubes and screens, but this use has been abandoned since the toxicity of the metal was recognised. It is now widely used in electronics, in ceramic chip carriers, resistor cores and laser tubes. Heat-resistant shields of rocket and re-entry cones, as well as crucibles are also made from the oxide.

BERYLLIUM IN FOOD AND BEVERAGES

There is a lack of reliable information on levels of beryllium in foods and diets. An early study, in the 1940s, before good analytical procedures had been developed for low levels of the element, estimated daily intake in an industrial society to be about 100 μg, with levels of about 0.1 mg/kg (dry weight) in a variety of vegetables and cereals[2]. In the light of later reports of extremely low levels of the metal in human tissue[3] these early reports of levels in the diet seem exaggerated. More recent data suggest that levels in food are in the nanogram-per-gram range, with daily intake about 12 μg from food[4]. However, Tsalev has commented that even these figures are approximate and need verification[5].

Some plants have been reported to accumulate beryllium and these could be a source of high intake in animals and humans. Cigarette smoke may also contain the metal. Since beryllium alloys are increasingly used in electrical switches and relays and components of instruments and computers, concern has been expressed that situations might arise in the food processing industry where beryllium-containing equipment was used in places and parts for which it was not originally intended. Because of the highly toxic nature of the element at very low levels of concentration, the consequences of such misuse for consumers could be considerable[6].

ABSORPTION AND BIOLOGICAL EFFECTS OF BERYLLIUM

Animal experiments indicate that absorption of beryllium from ingested food is low. The metal is carried in the bloodstream, apparently as a colloidal phosphate bound to protein. Some of the metal is incorporated into bone after initial retention in liver. Its behaviour in the body is akin to that of magnesium[7]. Most of the absorbed beryllium is excreted in urine. Some may appear in milk.

Beryllium and its compounds are highly toxic to animal cells, even at very low levels of concentration. The metal is a powerful inhibitor of several enzymes, in particular alkaline phosphatase. It also affects protein and nucleic acid metabolism. Beryllium interferes with normal immune function. Inhaled beryllium gives rise to an incapacitating lung disease and possibly cancer. As has been pointed out in a study of beryllium toxicity by Skilleter[8], 'the main hazards to humans occur almost exclusively in the occupational beryllium industries. The oral ingestion of beryllium compounds is of minimal concern'. However, some other authors referred to by Nielsen[9] believe that, like other trace elements given in high doses, ingested beryllium can be toxic.

The dangerously toxic properties of beryllium were first observed on a large scale during the 1940s when the term 'beryllosis' was given to the illness which followed rapidly on inhalation of dust or fumes of the then relatively unknown metal and its salts. Hundreds of workers, including many in the newly established nuclear industry, in which the metal and its products were beginning to be used in quantity, developed the illness and some died. There were also outbreaks of 'neighbourhood cases' of beryllosis, among people living in the neighbourhood of plants that used the metal. As has been observed earlier[1], it took some time for health authorities to recognise the source of the illness. Unfortunately this was not before beryllium had been used in the production of fluorescent tubes for the domestic market. Though this use has now been stopped, until well into the 1950s poisonings through contact with old fluorescent light tubes continued to be reported. At the present time most countries have placed severe restrictions on the use of beryllium and have set stringent occupational exposure standards for industries that must use the metal. Despite these restrictions, a small number of new cases of beryllium poisonings continue to occur each year.

ANALYSIS OF FOODSTUFFS FOR BERYLLIUM

As has been noted, very few reliable recent reports on levels of beryllium in foods or on methods for analysis of the metal have been published. Tsalev[5] has commented that graphite furnace atomic absorption spectrophotometry is a suitable method for the determination of nanogram-per-gram quantities of beryllium in biological samples. He recommends the use of an acid mixture for digestion.

A sensitive method for determination of the element is gas chromatography coupled with an electron capture detector or a mass spectrometer.

STRONTIUM

We have already looked briefly at the radioactive isotope of this element, ^{90}Sr, a potentially hazardous by-product of nuclear fission. Here we will consider non-radioactive strontium.

CHEMICAL AND PHYSICAL PROPERTIES

Strontium is element number 38 in the Periodic Table, with an atomic weight of 87.62. It is a silvery white, malleable and ductile metal, with a melting point of 768° C. As has been mentioned, it is one of the alkaline earth metals, with chemical properties similar to those of calcium.

PRODUCTION AND USES

Strontium is an abundant element in rocks, soil and water. It occurs in seawater at a concentration of about 8 mg/litre[10]. Its principal ores are celestine, $SrSO_4$, and strontianite, $SrCO_3$. The metal is extracted from these ores by electrolysis. World production is about 12 000 tonnes per annum.

Strontium metal is used in relatively small amounts, as a 'getter' to remove traces of gas from vacuum tubes, in the construction of permanent magnets and in certain iron alloys. The principal use of the element is in the form of its salts. Strontium nitrate, and other compounds, are used to give a bright crimson colour to fireworks and flares. They are also used in ceramics and plastics and, in small amounts, in certain pharmaceutical preparations. Strontium hydroxide is used in sugar refining as it combines with sucrose to form an insoluble saccharate.

STRONTIUM IN FOOD AND BEVERAGES

Few data are available in the literature on levels of non-radioactive strontium in foods. Nielsen[9] has summarised the available data. In general, foods of plant origin have higher levels of strontium than have animal-based foods. Levels of between < 1 to about 40 mg/kg have been reported in plants growing on normal, uncontaminated soil. In cereals strontium tends to concentrate in the bran and in root crops in the outer peel. Some plants apparently have the ability to accumulate the element to more than normal levels of concentration. Brazil nuts are believed to be a particularly rich source of strontium. Dairy products have been reported to contribute a major portion of daily strontium intake. This has been estimated to be no more than a few milligrams.

ABSORPTION AND METABOLISM OF STRONTIUM

The behaviour of strontium in the body is similar to that of calcium. Both elements are more readily absorbed by young than old animals, and absorption is increased by vitamin D, lactose and certain amino acids. Parathyroid hormone also plays a part in metabolism of both elements.

Absorbed strontium is retained in bone and teeth. There seems to be some competition between calcium and strontium for deposition in bone.

The ingestion of large amounts of strontium was found to inhibit calcification and stunt growth in young mice, giving rise to a condition known as 'strontium rickets'. There is also evidence that a chronically high intake of strontium, for example, from drinking water, may lead to dental caries in children[11].

There is no conclusive evidence that strontium is essential for man or other animals. Absence of the element from the diet of experimental animals, however, resulted in some growth depression. Strontium is also believed to be able to substitute for calcium in some enzyme systems, but whether it is essential or not for their function has not been established.

ANALYSIS OF FOODSTUFFS FOR STRONTIUM

Strontium is readily determined by atomic absorption spectrophotometry. At low concentrations it may be necessary to carry out a preconcentration step. Dry ashing or wet acid digestion may be used for sample preparation. However, sulphuric acid should not be used as it can cause losses due to sulphate formation[5].

BARIUM

Barium belongs, along with calcium and strontium, to what are known as the alkali metals. Its widespread distribution in the environment, and its use in many domestic and medicinal applications, make it of interest as a potential food contaminant. Though it is generally accepted that barium, like calcium, its fellow member of Group II of the Periodic Table, is a hazard-free and perhaps even an essential component of the diet, there is at least one report of serious poisoning caused by its presence in food and consequently it will be briefly considered here.

CHEMICAL AND PHYSICAL PROPERTIES

Barium is number 56 in the Periodic Table, with an atomic weight of 137.3. It is the heaviest of the alkali earths, with a density of 3.5 and a melting point of 710° C. It is a soft, silvery yellow metal. The metal tarnishes readily in air.

The chemistry of barium is in many respects similar to that of calcium. It is very reactive and spontaneously ignites on exposure to air. It is a divalent element and forms compounds with all ordinary anions. Barium sulphate is of particular interest as one of the most insoluble compounds known.

PRODUCTION AND USES

There are a number of ores of barium, the best known of which is heavy spar or barytes, $BaSO_4$. The ore is usually reduced with charcoal to the sulphide, which is then used to produce the various compounds of the element which are the forms in which barium is mainly used industrially. The pure metal is produced by electrolysis followed by vacuum distillation.

Barium has a number of important applications. A very small amount is used to make special alloys with aluminium, nickel and magnesium. These were formerly used in the construction of radio valves and are today used for bearings and other specialised purposes.

Most barium is, in fact, used as it occurs in barytes, without any treatment apart from crushing. Some 80 per cent of the world's production of barytes is used to make special mud for oil and gas drilling. Barium compounds are used in the production of glass, as fillers for paper, textiles and leather, in ceramics, television tubes, bricks, as pigments in paints, and dyes, as lubricating oil additives, and in a number of domestic products. These include cosmetics, insecticides and rodenticides, detergents and bleach. Barium hydroxide is used in refining sugar and vegetable oils. An important medical use of barium sulphate, which relies on its insolubility and its opaqueness to X-rays, is as the basis of the 'barium meal' in radiological investigations. Thus barium is very much part of the modern human environment and way of life, a fact of some significance when it is realised that its compounds are poisonous.

BARIUM IN FOOD AND BEVERAGES

It is somewhat surprising that, in spite of the fact that, as Tsalev has noted[5], the primary route of exposure to barium for the general population is usually via food, we have few data on levels of the element in individual foods or diets. Those that we have do not appear to be very reliable and range from 10 to several thousand mg/kg[12]. Nielsen[9] has summarised available information on levels of barium in food and diets. He reports that daily intake in the UK is estimated to be about 0.6 mg, with similar levels reported for the US. Little is known about levels in individual foods. The element is said to be associated with calcium in foods and thus dairy products might be expected to be a good source. It has been estimated that dietary intake of barium originates from four main sources in almost equal amounts: milk, flour, potatoes, and high barium-containing foods, such as nuts consumed in minor quantities[13]. Brazil nuts have been found to contain up to 3000 mg/kg of the metal, apricot kernels up to 33 mg/kg and pecans 14 mg/kg.

ABSORPTION AND METABOLISM OF BARIUM

Barium is relatively poorly absorbed, with between 1 and 30 per cent of its soluble salts entering the blood from ingested food. Excretion of ingested barium is rapid, mainly in faeces. There is some retention in bone and teeth. Barium can cross the placental barrier.

There is no convincing evidence that barium is an essential trace nutrient for humans or other animals. Early studies which indicated that the element was required for normal growth in rats have not been confirmed. Even though the Ba^{2+} ion is extremely toxic when absorbed, the relative insolubility of most barium salts makes the likelihood of poisoning by normal dietary intake remote. However, accidental and homicidal poisonings have occurred by the oral route. Ingestion of barium-containing household chemicals, such as rodenticides, depilatores, and insecticides, have resulted in cardiac arrest and death within a few hours in a few cases. An outbreak of food poisoning due to natural contamination by barium of locally mined table salt in China has been reported[14].

ANALYSIS OF FOODSTUFFS FOR BARIUM

Sample preparation for analysis of food for barium must avoid use of sulphuric acid to prevent precipitation of barium sulphate. Dry ashing, followed by HCl extraction, is suitable, as is wet digestion with acids other than sulphuric.

Graphite furnace atomic absorption spectrophotometry is suitable for very low levels of barium in samples, though problems with carbide formation and background can occur. Zeeman mode background correction has been found to overcome this latter problem.

THE OTHER ELEMENTS — SUMMING UP

Not all the remaining elements which will be treated briefly here are true metals or even metalloids. However, they all are industrially associated with metals, mainly in alloys, and, as such, perform many of the same roles in human affairs as do the 23 elements that have already been considered. All of these additional elements can be found in foods, some in appreciable amounts. We do not know a great deal about their concentrations in diets or their effects on human metabolism, but it is likely that this situation will change as investigation on the health significance of metal contaminants of foods expands. What will be done here is to 'tag' these elements as potential problems in foods and once again leave it to future editions to expand on details when, and if, further research indicates that fuller treatment of any of them is called for.

BORON

Boron (atomic weight 10.8; atomic number 5) is a very widely distributed and, from the point of view of agriculture in particular, a biologically important element. It is a brown, amorphous powder and occurs naturally as its sodium salt, borax. It is used for hardening steel and for producing enamels, glass and ceramics. Its ability to absorb neutrons makes it ideal for use in steel alloys for making control rods in nuclear reactors. Boron and borax are used in water treatment, fertilisers and pharmaceuticals.

Boron is a normal component of the diet, occurring in all food groups. Plants of vegetable origin, including fruits and nuts, are rich sources of boron, with up to 13 mg/kg. Fish and meat have the lowest levels, with between 0.16 and 0.36 mg/kg, and dairy products and cereals with about 1 mg/kg each[15]. Thus vegetarians may have a relatively high intake of the element, compared to meat eaters. Daily intakes have been estimated by various authors to range from about 2 to 20 mg.

Boron is rapidly and almost completely absorbed from food in the intestine. Its essentiality for animals, including humans, has not been established, though there is growing evidence that in its absence animals develop poorly and develop bone abnormalities. It is suggested[15] that there is a human requirement of 2 mg/day and that without an adequate dietary intake, osteoporosis may develop.

Boron has a low level of toxicity when administered orally. However, large doses have been shown to cause poisoning in experimental animals.

Accidental poisoning has occurred in humans who have swallowed boron-containing pharmaceutical and household products.

Boron is difficult to determine in food by AAS because of carbide formation and matrix effects. Adequate background correction is essential. NAA has been used effectively on blood samples[16].

BISMUTH

Bismuth (atomic number 83; atomic weight 208.98) is produced mainly as a by-product of the refining of lead and copper. The metal is used as a low-melting point alloy for fuses, solders and in nuclear reactors. It has a number of other industrial applications, such as the manufacture of paints and rubber vulcanising. Many cosmetic and pharmaceutical preparations, such as face powders, antiseptics, absorbents and antacids use bismuth compounds.

Levels of bismuth in food are low, and normal dietary intake has been estimated to be about 5 μg/day[17]. Misuse of bismuth-containing pharmaceutical products may result in considerably increased intake of the metal. Bismuth and its compounds are not readily absorbed. Excretion is rapid, mainly in urine. There is no evidence that bismuth is an essential nutrient for man. Excessive intakes of water-soluble bismuth compounds can lead to renal damage and encephalopathy, as well as less serious problems of skin irritation and pigmentation[5]. Cases of poisoning of children who have accidentally consumed large amounts of bismuth-containing pharmaceuticals have occurred[18]. There are no reports of poisoning due to bismuth in normal foodstuffs.

GERMANIUM

Germanium was once considered a rarity and of no industrial significance. That opinion has changed in recent years. Germanium (atomic weight 72.59; atomic number 32) has the unique property of permitting an electric current to pass in only one direction. This rectifying power makes it invaluable in the electronics and allied industries for making semiconductors and transistors which have replaced valves for the amplification of minute electric currents. The metal is also used in the manufacture of special optical glasses. Germanium is obtained as a by-product of zinc refining and is also recovered from waste products of the coal and coke industry.

Germanium is reported to be widely distributed in foods but at very low levels, with a daily intake of less than 0.5 mg[19]. Absorption of the metal from ingested food is believed to be rapid and complete, followed by rapid excretion mainly in urine.

While in experimental animals germanium has been shown to produce toxic effects, including degenerative changes in liver and kidney, the metal and its compounds are believed to be of low toxicity to humans, especially at the levels normally ingested in foods. However, germanium dioxide is toxic to other organisms, including bacteria[20].

The determination of germanium in foods and other biological materials is difficult. Very low concentrations are normally present and even with flameless AAS, there is insufficient sensitivity. Hydride generation techniques, using germane (GeH_4) and plasma source AES are recommended by Tsalev[5].

LITHIUM

Lithium is one of the alkali metals, along with sodium, potassium, rubidium and caesium. It is the lightest of the group, in fact the lightest solid substance known. Its specific gravity is 0.57, atomic weight 6.94 and atomic number 3. It is a silvery white metal. Chemically it is similar to sodium, though less active.

Its special interest in the context of this book is its use in psychiatric medicine to treat manic depressive psychosis. Its pharmacological effects are to alter the conductivity of nerves. Toxicity from misuse of lithium medication is well known, though the same has not been reported from ingestion of lithium in foods. There is some evidence that lithium may pass in breast milk to infants of mothers being treated for depression[21].

Levels of lithium in foods are believed to be low, but data are lacking on dietary intake. Hard water may contain significant amounts of the element. It has been suggested that there is an inverse relation between heart disease and levels of lithium in water in certain areas.

Both flame and electrothermal AAS are suitable for the determination of lithium in biological samples after dry or wet ashing.

TELLURIUM

Tellurium (atomic weight 127.6, atomic number 52) is a relatively rare metal of the sulphur family, very similar to selenium in chemical properties. The metal is extracted as a by-product of the production of copper and some other metals. It has a number of industrial uses, mainly in metallurgy. It is used to improve the properties of steel and other alloys for special purposes. It is also used in the manufacture of glass, plastics and rubber. Small amounts are used as catalysts in the chemical industry, in explosives, antioxidants and the electronics industry.

Information on levels of tellurium in individual foods is largely lacking, but it is believed to occur in most foods at low levels. Daily intake has been estimated to be about 100 μg. This may increase if food has been stored in metal containers[22]. About 10–20 per cent of ingested tellurium appears to be absorbed by the body, followed by rapid excretion both in urine and faeces. Some accumulation appears to occur in the skeleton.

The toxicity of orally administered tellurium is low. A peculiar effect of such ingestion is the formation of dimethylfluoride which is exhaled from the lungs. This gives the characteristic garlic smell noted in animals and man following exposure to tellurium and its compounds.

Serious metabolic effects have resulted from long-term exposure of experimental animals to tellurium compounds. Liver damage and impairment of nervous function resulted from the exposure. Similar, though less obvious symptoms have been observed in workers exposed to the metal. Accidental oral ingestion of tellurium compounds have caused a number of human fatalities[23].

Tsalev[5] has noted that serious methodological problems, including analyte loss, preconcentration difficulties, background and matrix effects, still remain to be solved with regard to the determination of tellurium. However, both graphite furnace and hydride-generation AAS are potentially applicable to the problem.

THALLIUM

Thallium has come to public notice in recent years because of a number of serious poisoning incidents that have resulted from the misuse of thallium-containing pesticides. An editorial in the *Lancet* called for the utmost vigilance in the use of such products and recommended that restrictions be placed on their availability[24]. Articles in newspapers and magazines brought the problem to the attention of the community and alerted them to the dangers of what, until then, had been a relatively unknown metal.

Thallium (atomic number 81; atomic weight 204.4) is a white, malleable metal resembling lead, which is widely distributed in rocks and soils in small amounts. It is usually recovered as a by-product in the smelting of lead and zinc ores. It has a number of industrial uses, in alloys, as a catalyst in the mineral oil industry, in electronics, making of pigments and special glass, and, of particular interest in the present context, in pesticides, especially rodenticides.

Little is known about levels of thallium in foods and diets. Levels in diets are believed to be very low, perhaps about 2 μg/day[5], but contamination with thallium-containing pesticides has been known to increase intake to serious levels[25]. Thallium contamination of the environment, and, potentially, of the diet, can be caused by smoke from coal-burning power stations, smelting of metals, glass manufacture and certain agricultural fertilisers. Some plants have been reported to accumulate high levels of thallium naturally from soil and can be toxic to cattle[9].

Thallium is readily and totally absorbed from food. The metal also is believed to accumulate in the body with age. It is a highly toxic element[26], which primarily affects the central nervous system. Acute poisoning results rapidly in nausea, diarrhoea and abdominal pain. Delirium, convulsions and circulation problems follow and may culminate in death. If recovery results, it may be incomplete, leaving mental abnormalities, as is often the case with children.

ZIRCONIUM

Zirconium (atomic number 40; atomic weight 91.21) is included among those elements known as the 'rare earths', and yet is widely distributed in the earth's crust. It is, in fact, more abundant in soils than are copper and zinc. It has many industrial uses, especially in alloys. Its oxide, zirconia, has excellent refractory qualities and is extensively used for this purpose. The metal and its compounds are used as industrial abrasives and in flame proofing.

Little information is available on levels of the element in foods and diets. It is believed to be present in most foods at low levels of concentrations. Daily intake has been calculated to be about 3.5 mg[27]. Little is known about its absorption and metabolism. In rats zirconium has a low level of toxicity when ingested. There is no indication that zirconium in the normal human diet is a cause of concern.

THE REMAINING METALS

There are still 50 elements that could warrant inclusion in this book. Several of them, such as gold, indium, niobium and rubidium have been considered appropriate for inclusion in their reviews of important metals in foods and the environment by Nielsen[9] and Tsalev[5], and could have been discussed here. However, at the present time we still know very little about these other metals as contaminants of food and the diet. They are still mainly of interest in the area of occupational health. That situation will undoubtedly change as many of these metals find new and extended industrial and domestic use. As a result they may, like lead and other metals discussed here at length, become food contaminants of note, also deserving attention in a study such as this.

One major factor that will help to bring these other metals to the attention of food scientists and others interested in metal components of the diet, will be improved analytical techniques. We have already seen how that has affected our knowledge of such metals as selenium and chromium. Of major significance in this regard have been improvements in atomic absorption spectrophotometry, especially through the ability to overcome background interference, as well as the introduction of inductively coupled plasma emission photometry. Accompanying the improvements in analytical techniques have been a better appraisal of the intricacies involved in trace metal analyses of biological samples, and, of no less importance, the availability and routine use by many analysts of certified reference materials[16]. It is to be expected that over the next decade, advances will be made in our understanding of the metals we already know a good deal about as well as of those still little understood.

Not as much progress has been made in one area of this study as was expected when the first edition of this book was published in 1980. Then it seemed that the next important step in understanding metal contamination of foods was about to be taken and rapid progress could be expected. This was with regard to the speciation of elements in food, and the need to go beyond the simple investigation of levels of total metals in food. While some progress has been made in that direction, in particular with regard to mercury, lead and to a lesser extent, selenium, we are still very ignorant of the chemical forms of metals in foods and their physiological significance. We still, for instance, report total iron content of foods in tables of food composition, though we know that it makes a very great difference from the point of view of nutrition, whether the iron in a meal is in the Fe(II), Fe(III) or organically-bound form.

It is to be hoped, however, that the compilation of data on the more important metal contaminants of food presented here, professionally or for any other reason, will be of use to the food scientist, nutritionist and others who are concerned with food and what it contains. This synthesis of information, some obtained by the investigations of the author and his colleagues and students over many years, with the bulk gleaned from worldwide scientific literature, will, it is hoped, serve as a helpful stepping stone for the further advancement of our knowledge of metal contaminants of our diets.

REFERENCES

1. HYSLOP, F. (1943). *NATIONAL INSTITUTE OF HEALTH BULLETIN NO.* 181 (US Government Printing Office, Washington, DC).
2. EISENBUD, M. (1949). Beryllium in the environment. *J. Ind. Hyg. Toxicol.*, 31, 282–94.
3. SCHUBERT, J. (1958). Beryllium and berylliosis. *Scientific American*, August, 322–7.
4. STIEFEL, T., SCHULZE, K., ZORN, H. *et al.* (1980). Toxicokinetic and toxicodynamic studies of beryllium. *Arch. Toxicol.*, 45, 81–9.
5. TSALEV, D.L. (1983). *Atomic Absorption Spectrometry in Occupational and Environmental Health Practice*, Vol. I, 96 (CRC Press, Boca Raton, Florida).
6. WHITMAN, W.E. (1978). Heavy metals in foods. *Proc. IFST (UK)*, 11, 86–90.
7. REEVES, A.L. (1979). Beryllium, in *Handbook on the Toxicology of Metals*, eds FRIBERG, L., NORDBERG, G.F. and VOUK, V.B. (Elsevier/North Holland, Amsterdam).
8. SKILLETER, D.N. (1990). To Be or not to Be — the story of beryllium toxicity. *Chemistry in Britain*, January 1990, 26–30.
9. NIELSEN, F.H. (1986). in *Trace Elements in Human and Animal Nutrition*, ed. MERTZ, W., 449 (Academic Press, New York and London).
10. BOWEN, H.J.M. (1966). *Trace Elements in Biochemistry* (Academic Press, New York).
11. ANON. (1978). Strontium, other trace elements and dental caries. *Nutr. Revs.*, 36, 334–7.
12. BEESON, K.C. (1941). *Misc. Publications, US Dept. Agric.*, 369, 1–164.
13. REEVES, A.L. (1979). Barium, in *Handbook on the Toxicology of Metals*, eds FRIBERG, L., NORDBERG, G.F. and VOUK, V.B. (Elsevier/North Holland, Amsterdam).

14. ROZA, O. and BERMAN, L.B. (1971). Barium poisoning due to consumption of contaminated salt. *J. Pharmacol. Exp. Ther.*, 177, 433–9.
15. NIELSEN, F.H. (1988). Boron — an overlooked element of potential nutritional importance. *Nutrition Today*, January/February, 4–7.
16. IYENGAR, V. and WOITTIEZ, J. (1988). Trace elements in human clinical specimens: evaluation of literature data to identify reference values. *Clin. Chem.*, 34, 474–81.
17. BOITEAU, H.L., CLER, J.M., MATHE, J.F. *et al.* (1976). Relationship between the course of bismuth encephalopathy and blood and urine bismuth levels. *Eur. J. Toxicol. Environ. Hyg.*, 9, 233–7.
18. BOYETTE, D.P. (1946). Bismuth poisoning of children by pharmaceutical compounds. *J. Paediat.*, 28, 193–7.
19. HAMILTON, E.I. (1980). The need for trace element analyses of biological materials in the environmental sciences, in *Elemental Analysis of Biological Materials* (International Atomic Energy Agency, Vienna).
20. VAN DYKE, M.S., LEE, H. and TREVORS, J.T. (1989). Germanium toxicity in selected bacterial and yeast strains. *J. Indust. Microbiol.*, 4, 299–306.
21. IYENGAR, G.V. (1980). *Elemental Composition of Human and Animal Milk.* TEC-DOC−269 (International Atomic Energy Agency, Vienna).
22. GLOVER, J.R. and VOUK, V. (1979). Tellurium, in *Handbook on the Toxicology of Metals*, eds FRIBERG, L., NORDBERGY, G.F. and VOUK, V. (Elsevier/North Holland, Amsterdam).
23. KEAL, J.H.H., MARTIN, N.H. and TUNBRIDGE, R.E. (1946). Tellurium toxicity. *Br. J. Indust. Med.*, 3, 175–6.
24. ANON. (1970). Thallium poisoning. *Lancet*, ii, 564–5.
25. REED, D. (1963). Thallium intoxication. *J. Am. Med. Assoc.*, 183, 516–22.
26. SPIRIDONOVA, V.S. and SHABALINA, L.P. (1977). Toxicology of thallium and its compounds, in *Problems of Occupational Hygiene*, ed. TARASENKO, N.J., 68–79 (VNIIMI-MZ-USSR, Moscow).
27. SCHROEDER, H.A. and BALASSA, J.J. (1966). Trace elements in food: zirconium. *J. Chron. Dis.*, 19, 537–42.

INDEX

271